AF824759

# PUZZLES

## INVOLVING POLYCUBES, TRIANGULATIONS AND POLYHEXES

RAYMOND R. FLETCHER III

Copyright © 2021 by Raymond R. Fletcher III.

All rights reserved. No part of this publication may be reproduced, distributed, or transmitted in any form or by any means, including photocopying, recording, or other electronic or mechanical methods, without the prior written permission of the author, except in the case of brief quotations embodied in critical reviews and certain other noncommercial uses permitted by copyright law.

The views expressed in this work are solely those of the author and do not necessarily reflect the views of the publisher, and the publisher hereby disclaims any responsibility for them.

Printed in the United States of America.

Library of Congress Control Number: 2021948025

| | | |
|---|---|---|
| ISBN | Paperback | 978-1-68536-093-1 |
| | eBook | 978-1-68536-094-8 |

**Westwood Books Publishing LLC**
Atlanta Financial Center
3343 Peachtree Rd NE Ste 145-725
Atlanta, GA 30326

www.westwoodbookspublishing.com

# CONTENTS

## SECTION III. POLYHEXES

# Introduction

A *polycube* is a collection of unit cells or cubes which are joined face to face. Each cube has 8 points (or vertices), 6 faces and 12 edges. The object of a polycube puzzle is to label each point with *numbers* so that the *sum* of the 4 numbers on each face of every cube of the polycube is the same. The numbers consist of elements from the set of *integers (mod) n* denoted $Z_{n,}$ and sums are computed *mod(n)*. For example $Z_6$ = (0,1,2,3,4,5). To add mod 6 simply add the two integers in the usual way then divide by 6 and take the remainder. The addition table for $Z_6$ is given by:

| + | 0 | 1 | 2 | 3 | 4 | 5 |
|---|---|---|---|---|---|---|
| 0 | 0 | 1 | 2 | 3 | 4 | 5 |
| 1 | 1 | 2 | 3 | 4 | 5 | 0 |
| 2 | 2 | 3 | 4 | 5 | 0 | 1 |
| 3 | 3 | 4 | 5 | 0 | 1 | 2 |
| 4 | 4 | 5 | 0 | 1 | 2 | 3 |
| 5 | 5 | 0 | 1 | 2 | 3 | 4 |

The set of elements $Z_n$ with the addition just described is the simplest example of an important mathematical object in modern algebra known as an *abelian group.* A *group* is a set of elements *closed* under an associative binary operation. In addition, a group is required to have an *identity element* and every element is required to have an *inverse.* An abelian group has the additional property that the binary operation is commutative. For $Z_n$ The binary operation is addition (mod n) illustrated for n = 6 in the above table. The elements of $Z_n$ are usually represented by the set of integers {0, 1, 2, …, n-1}. Addition (mod n) of two elements of $Z_n$ results in another element of $Z_n$. This is the required closure property. The element 0 is the identity since x + 0 = x for any element x in $Z_n$. Since x+y = y+x for any two elements of $Z_n$, the addition is commutative and since x+(y+z) = (x+y)+z for any three elements x,y,z of $Z_n$, thc addition is associative. For each element x there is an element -x with the property x + (-x) = 0. Thus the requirement that every element have an inverse is satisfied. In the addition table for $Z_6$ illustrated above, elements 1, 5 are inverses; 2, 4 are inverses and the elements 0, 3 are their own inverses. Thus 5 = -1; 4 = -2; 3 = -3 and 0 = -0. The elements of $Z_n$ can be represented using this notation, e.g., $Z_6$ = {0, 1,-1,2,-2, 3} or $Z_7$ = {0, 1, -1, 2,-2, 3,-3}. This representation is often quite useful in solving the puzzles; in fact it is used exclusively to represent the solutions to the polyhex puzzles (Section III).

Solving polycube puzzles is a fun way to become acquainted with some of the basic properties of $Z_n$. In Figure 1 we illustrate a simple polycube with 12 points and a solution over $Z_{12}$. It is required that each of the 12 elements of $Z_{12}$ occur among the labels assigned to the vertices of the polycube.

Figure 1: polycube with solution over $Z_{12}$

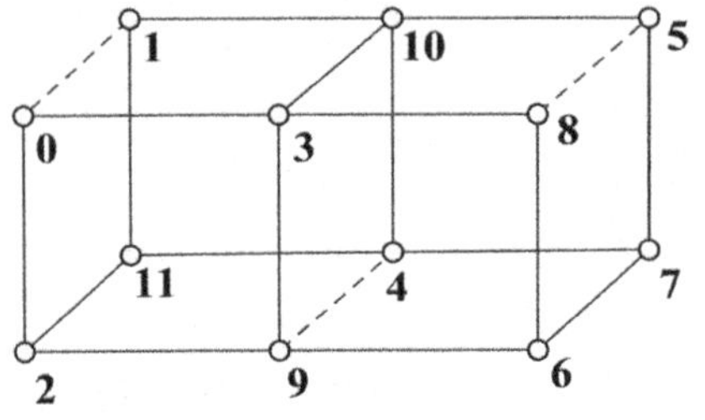


Note that the sum of the 4 elements on each face is 14 or 26 both of which are equivalent to 2 (mod 12) since 2 is the remainder when either number is divided by 12. We say the *face sum* is F = 2. Unlike Sudoku puzzles there are many correct solutions possible. For example, any element of the group $Z_{12}$ can be added to each point in Figure 1 to obtain a new solution with possibly a different face sum. If 1 is added to each point in Figure 1 we obtain the solution given in Figure 2 which now has face sum F = 6. We call the solutions given in Figures 1, 2 *equivalent.*

Figure 2: New solution

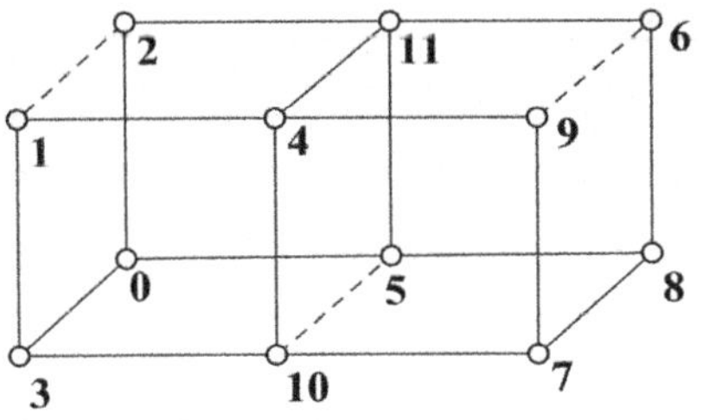


One important clue to finding a solution is that edges which are opposite each other in any cube must have the same sum. Thus each of the dashed bars in Figures 1, 2 must have the same sum (mod 12). Each cube has 4 pairs of opposite edges with the same sum. In fact, if all such pairs of edges have the same sum then all faces must have the same sum, and thus a solution is obtained. A more complicated example is given in Figure 3.

Figure 3: Polycube over $Z_{28}$

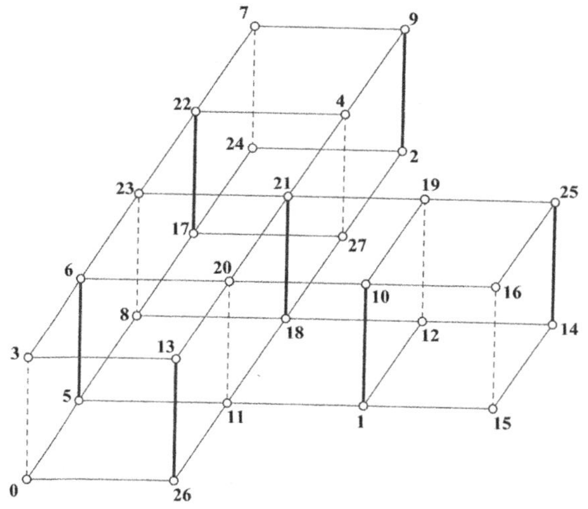


In this example the dashed edges must all have the same sum (mod 28); in this case the sum is 3. Similarly the bold edges must all have the same sum (mod 28) which is 11. All polycube puzzles are included in Section I.

The puzzles in Section II involve triangles. These fall into two categories: *regular polygonal triangulations (RPT)* and *irregular polygonal triangulations (IRT)*. An RPT consists of equilateral triangles which are joined edge to edge and an IRT consists of scalene triangles joined edge to edge. In either case the boundary of the figure is a polygon. **For these puzzles the edges, (instead of the vertices), are to be labeled with elements from $Z_n$ such that the three edges belonging to each component triangle have the same sum.** If the puzzle has n edges, then the full set of elements from $Z_n$ must be used to label the edges. In Figure 4 is illustrated an RPT with a solution over $Z_{11}$ and triangle sum 5 (mod 11), and Figure 5 depicts an IRT with a solution over $Z_{10}$ and triangle sum 6 (mod 10).

Figure 4: RPT with solution over $Z_{11}$

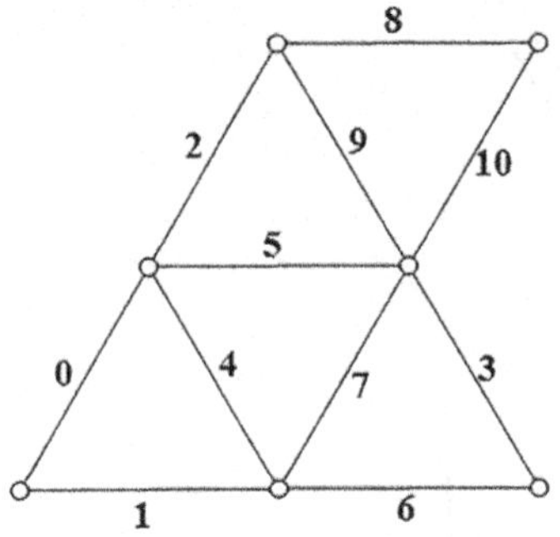


Figure 5: IRT with solution over $Z_{10}$

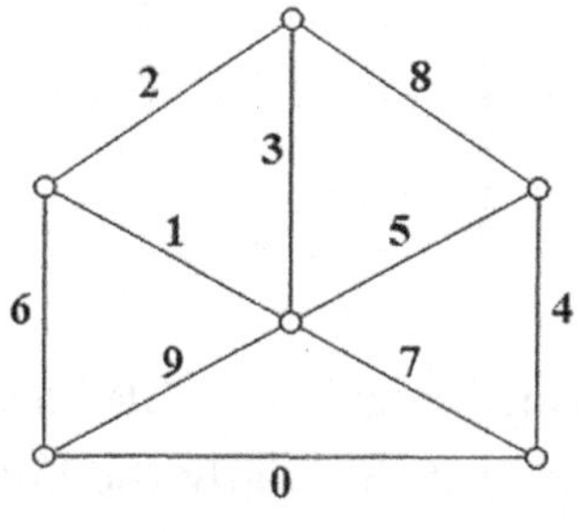

In Section III we have included puzzles involving hexagons but which have a similar idea for their solution. These collections of hexagons are called *polyhexes*. If a polyhex has n points then the object is to label each point with the elements from $Z_n$ such that the six points belonging to each component hexagon have the same sum (mod n). In Figure 6 we provide an example of a polyhex with 10 points with a solution over $Z_{10}$.

In Figure 6 the common hexagon sum is F = 5 (mod 10). On the right side the same solution is given in which $Z_{10}$ is represented using inverses.

Figure 6: Polyhex with solution over $Z_{10}$

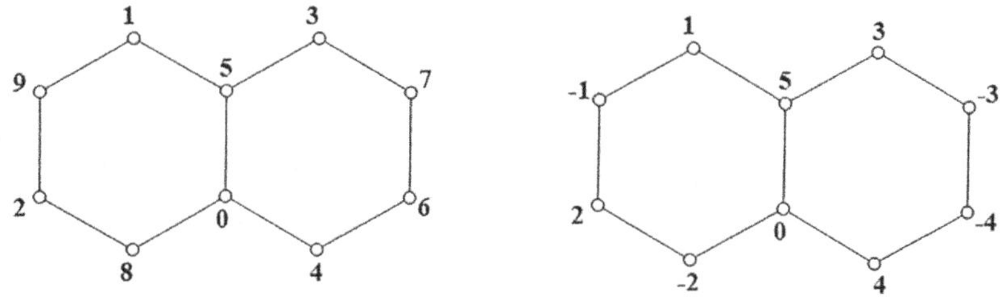


In some puzzles the points belonging to one face, (square, triangle or hexagon) have been assigned elements from the base group and the required common face sum F is given. In other puzzles the face sum is left open. In the back of the book solutions are given. This is only to show that a solution does exist; there are many solutions and the solution you find may be perfectly valid. Exercises and research questions are provided at the end of each Section with the latter marked with an (*). We encourage the creation and solution of the unlimited number of polycubes, polyhexes and triangulations not included. All that is needed, (after becoming familiar with addition (mod n)), is a couple of sharp pencils, a good eraser, some blank paper and a lot of ingenuity.

**Hints:** (The reader may prefer to skip this paragraph in order to find solution strategies for him or herself)

1. Instead of aiming to solve a particular polycube, triangulation or polyhex, it is better to create a puzzle along with a solution and then try to find it in the book. If not included among the puzzles in the book, then you have created and solved your own puzzle!

2. A *monotone solution* for a triangulation is one in which the common sum is the same regardless of the modulus. For example, in a triangulation over $Z_9$, the sums may be 3, 12 or 21 all of which are equal to 3 (mod 9). A solution can be found, (and is given in the solution section for puzzle 106), in which every sum is 12. This is an example of a monotone solution. It is often the case that one or two edges can be inserted in a monotone solution to create a solution to a larger puzzle.

3. Once a solution to a particular puzzle is obtained, look for ways to cut and reassemble the pieces to obtain a solution to a different puzzle.

SECTION I

# POLYCUBES

# 1. Linear Polycubes

Puzzle 1: Base group: $Z_{16}$; F = 6

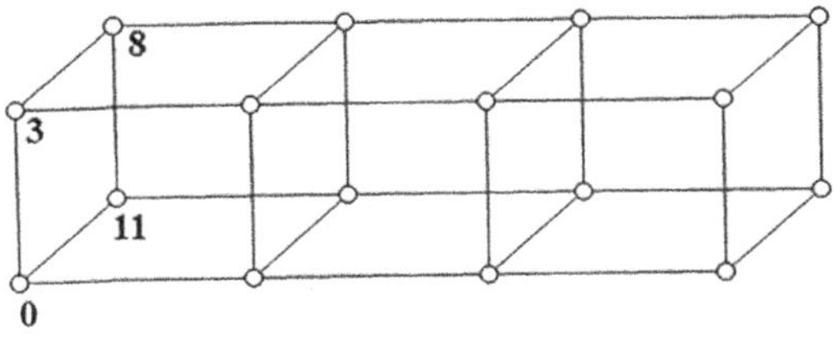


Puzzle 2: Base group: $Z_{20}$ ; F = 18

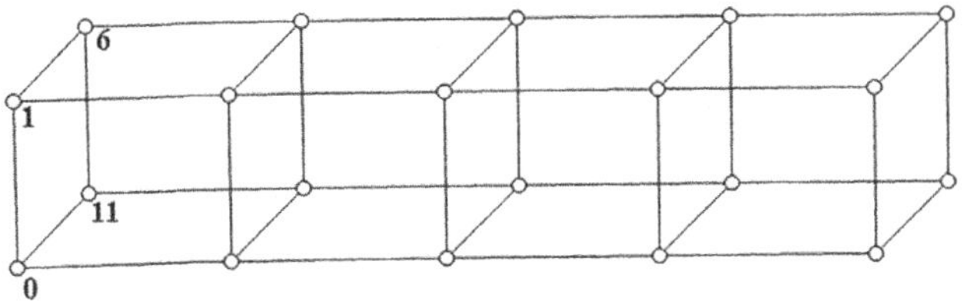


Puzzle 3: Base group: $Z_{20}$; F = 18

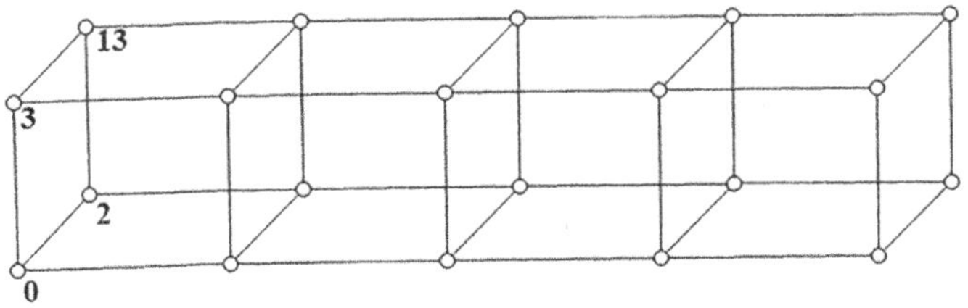


Puzzle 4: Base group: $Z_{24}$; F = 14

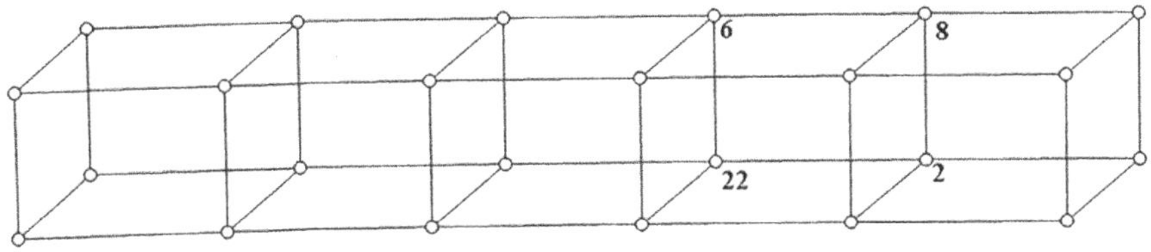

Puzzle 5: Base group: $Z_{28}$; F = 26

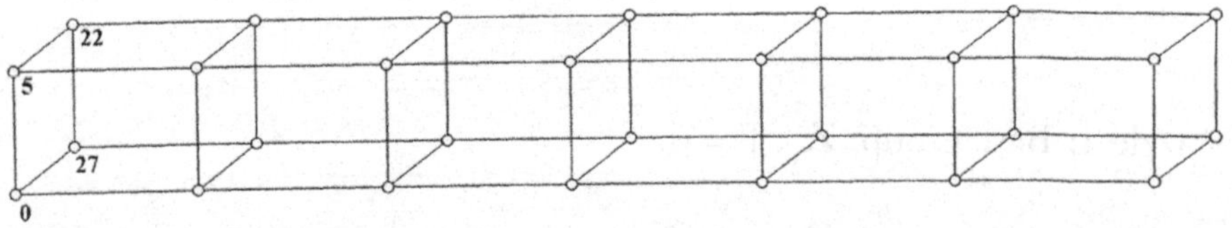


Puzzle 6: Base group: $Z_{28}$; F = 22

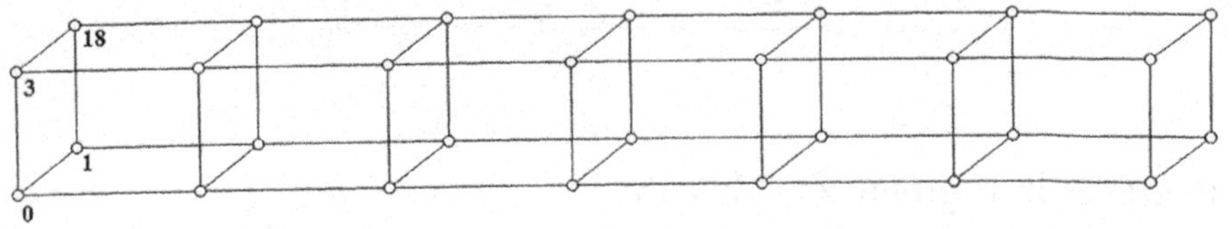

# 2. Base $Z_{16}$ and $Z_{20}$

Puzzle 7: Base group: $Z_{16}$

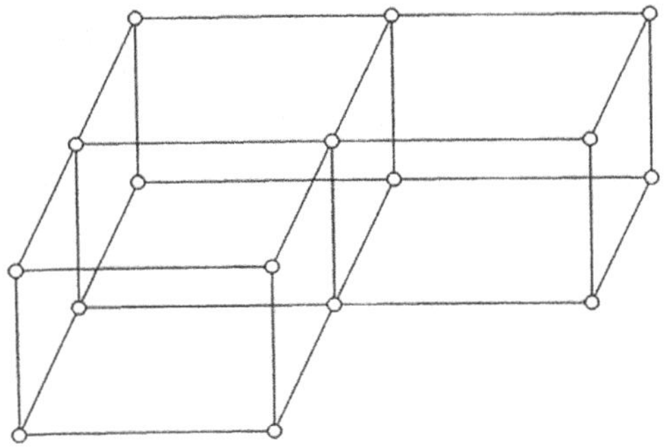

Puzzle 8: Base group: $Z_{20}$

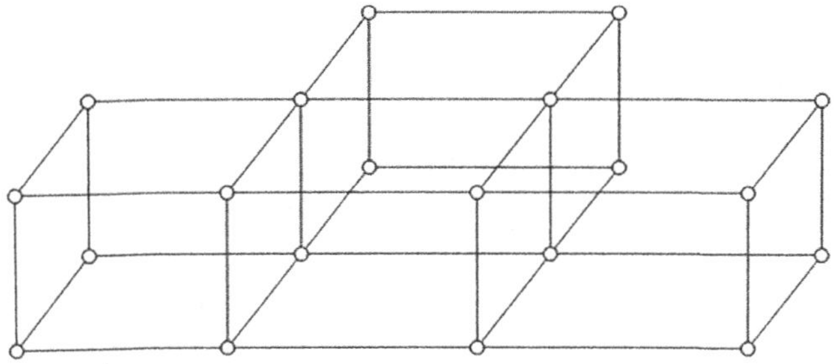

Puzzle 9: Base group: $Z_{20}$

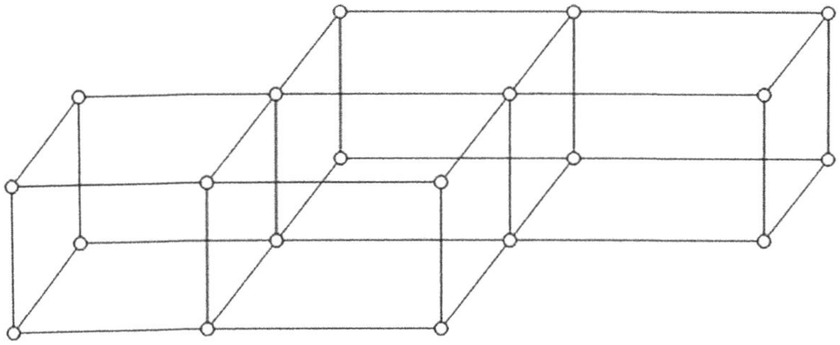

Puzzle 10: Base group: $Z_{20}$

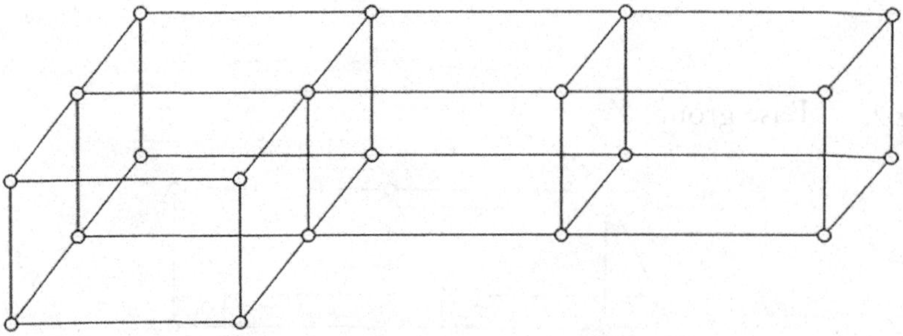

Puzzle 11: Base group: $Z_{20}$

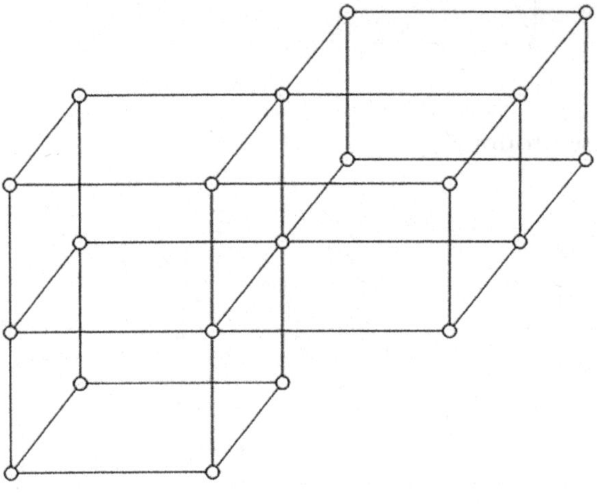

# 3. Base $Z_{24}$

Puzzle 12: Base group: $Z_{24}$

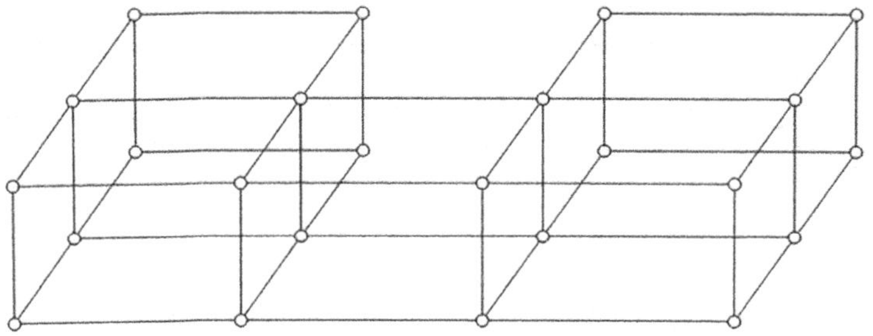

Puzzle 13: Base group: $Z_{24}$

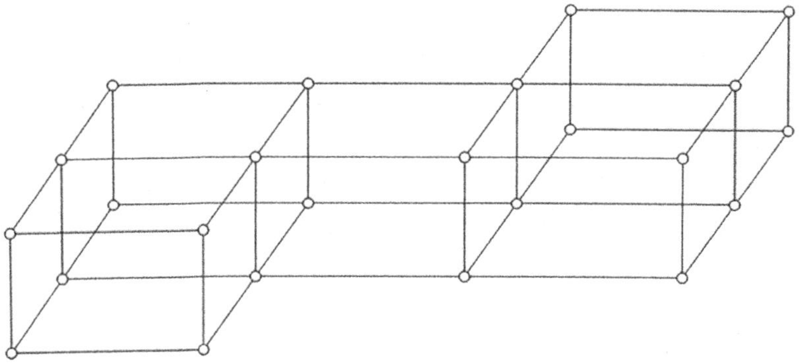

Puzzle 14: Base group: $Z_{24}$

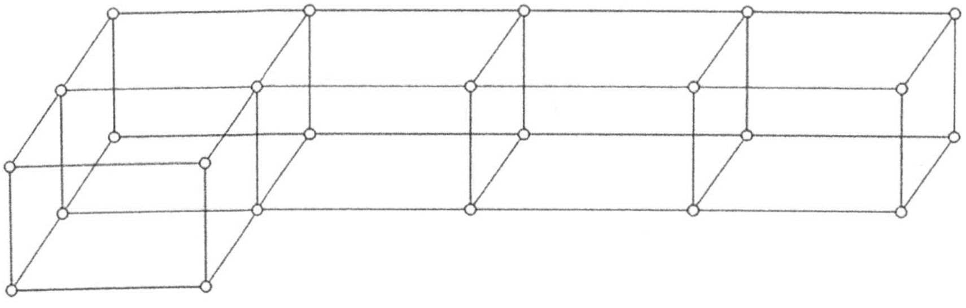

Puzzle 15: Base group: $Z_{24}$

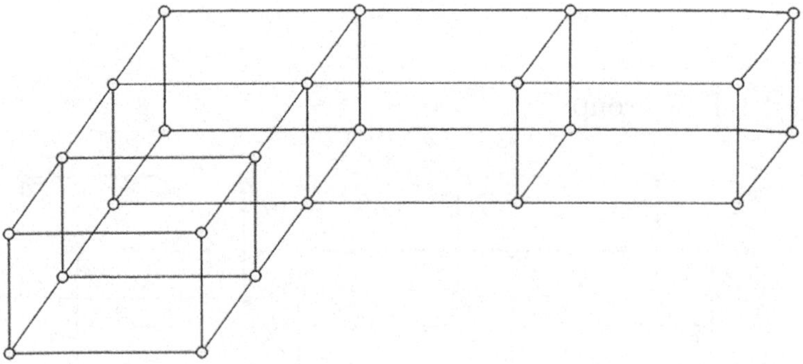

Puzzle 16: Base group: $Z_{24}$

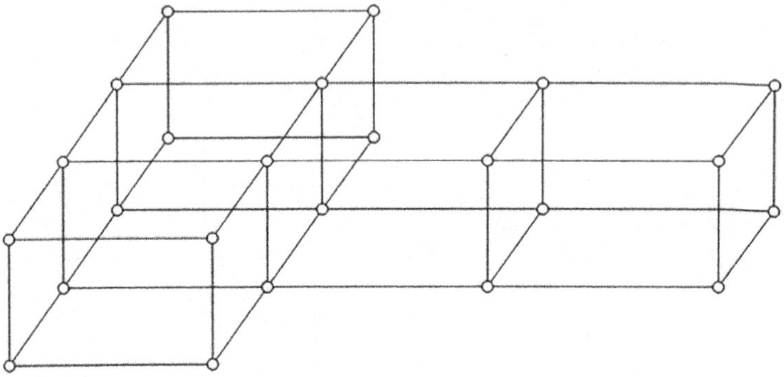

Puzzle 17: Base group: $Z_{24}$

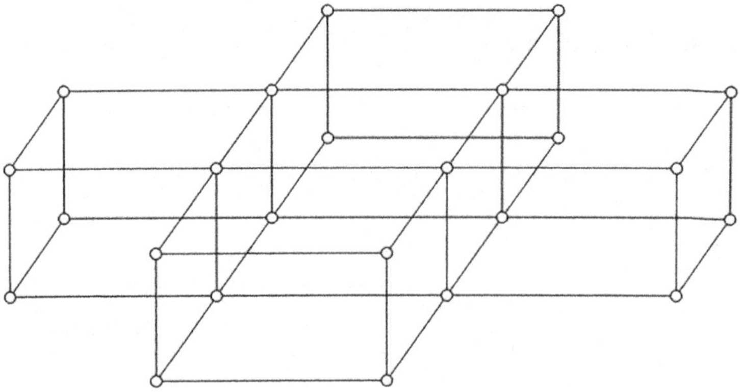

Puzzle 18: Base group: $Z_{24}$

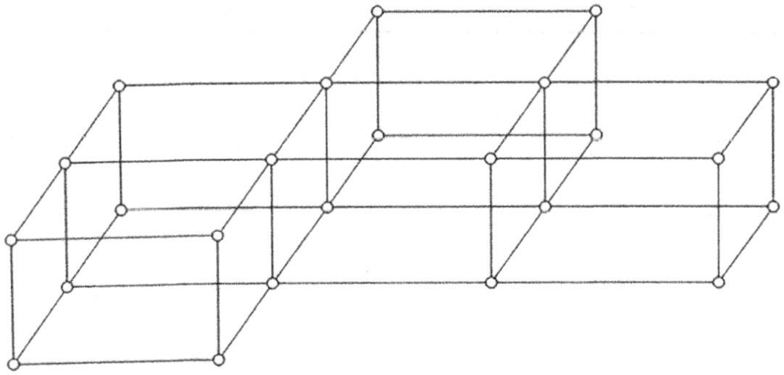

Puzzle 19: Base group: $Z_{24}$

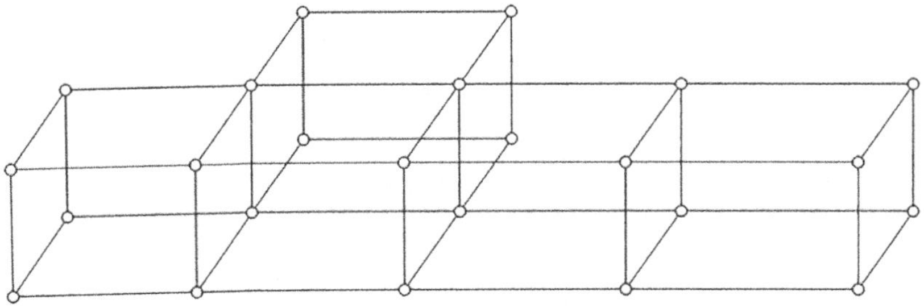

Puzzle 20: Base group: $Z_{24}$

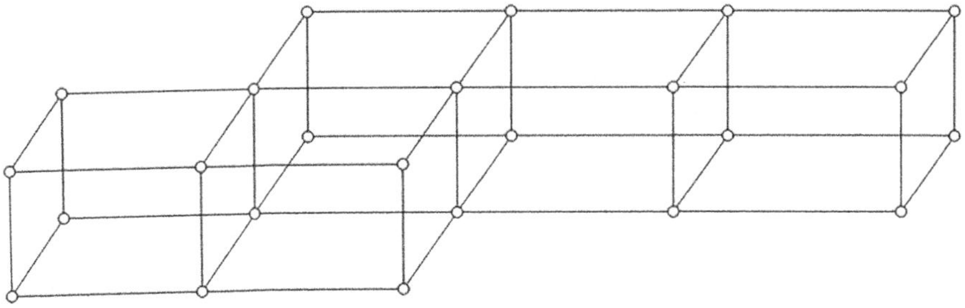

Puzzle 21: Base group: $Z_{24}$

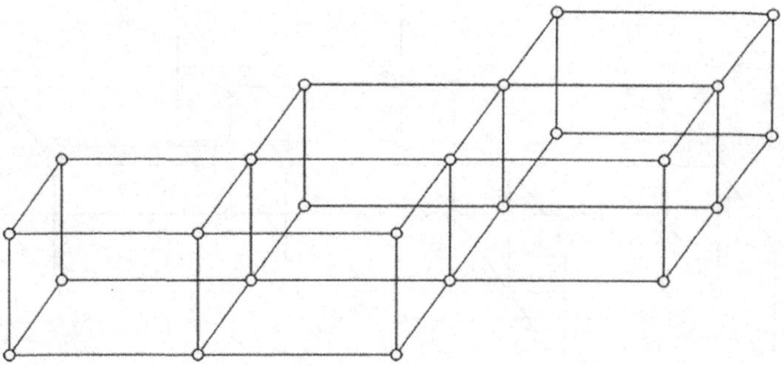

Puzzle 22: Base group $Z_{24}$ ; F = 14

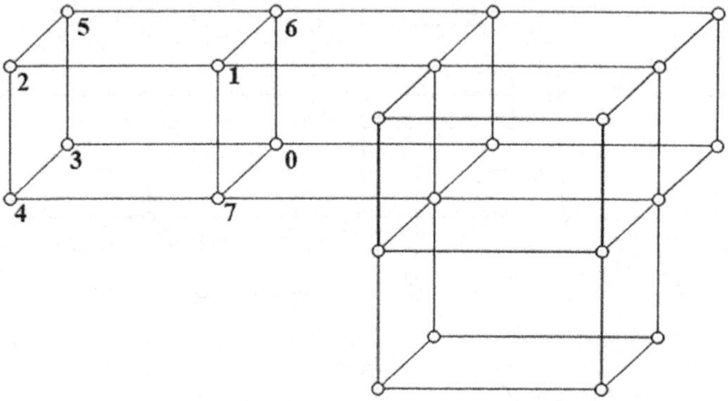


Puzzle 23: Base group $Z_{24}$ ; F = 14

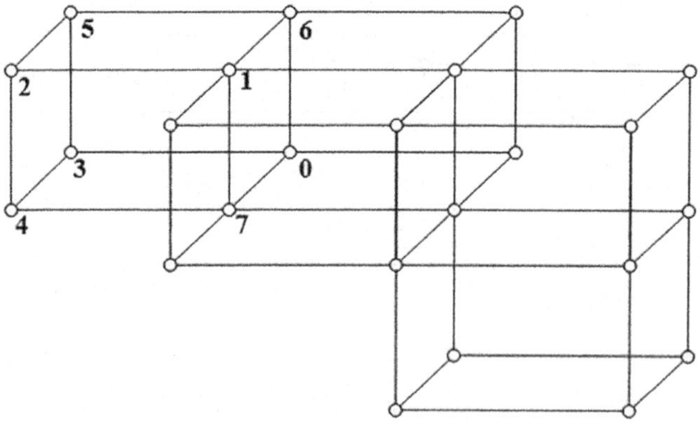

Puzzle 24: Base group $Z_{24}$ ; F = 14

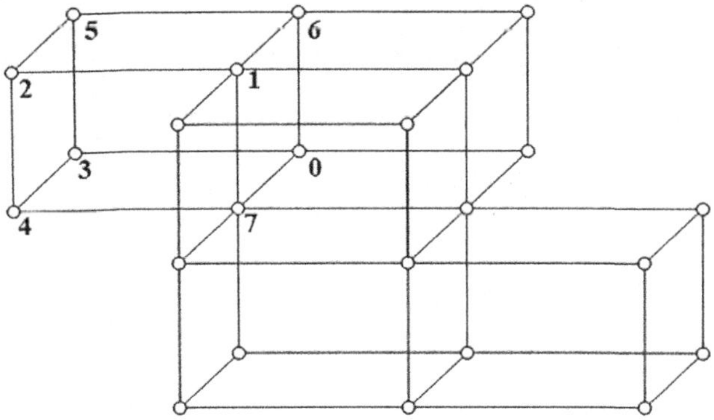


Puzzle 25: Base group $Z_{24}$ ; F = 14

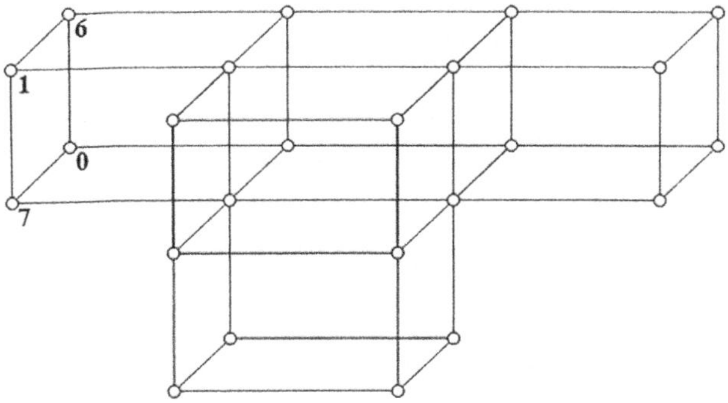


Puzzle 26: Base group $Z_{24}$ ; F = 14

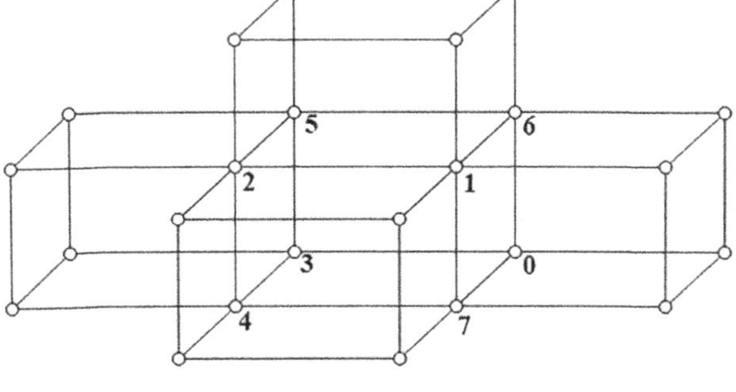

Puzzle 27: Base group $Z_{24}$ ; F = 14

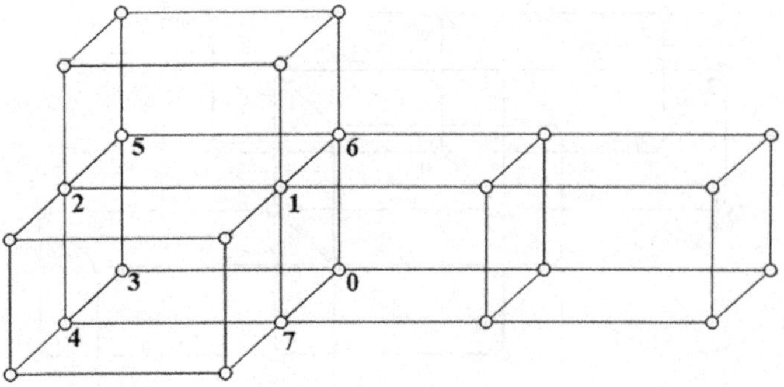


Puzzle 28: Base group $Z_{24}$ ; F = 10

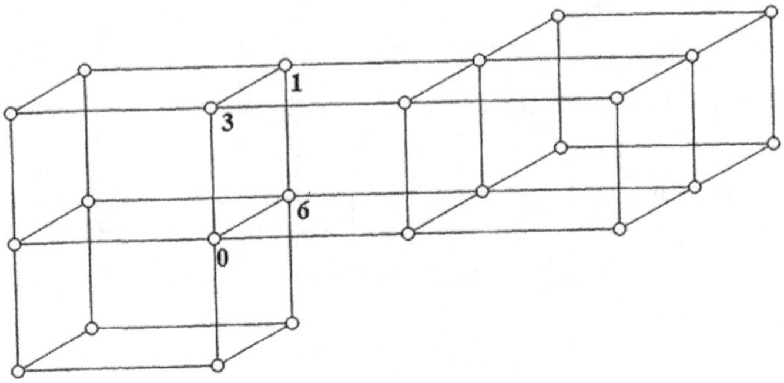


Puzzle 29: Base group $Z_{24}$ ; F = 10

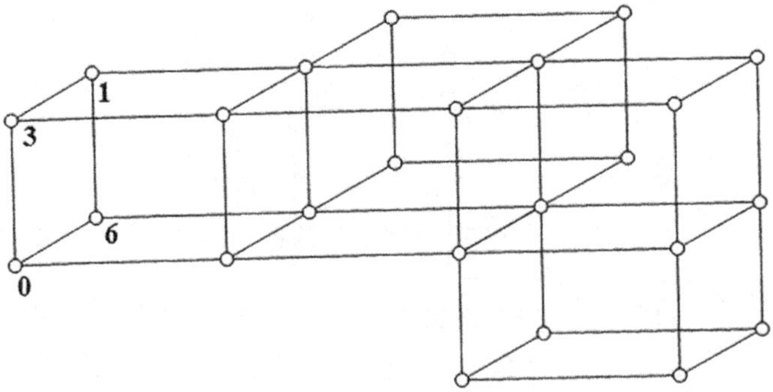

# 4. Base group $Z_{28}$

Puzzle 30: Base group $Z_{28}$; F = 14

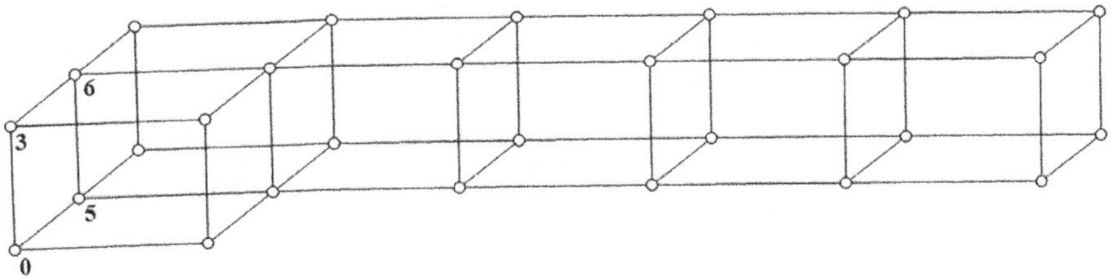


Puzzle 31: Base group $Z_{28}$ ; F = 14

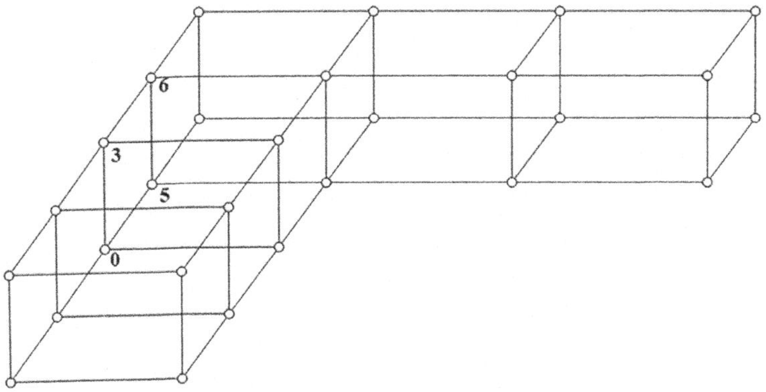


Puzzle 32: Base group $Z_{28}$; F = 14

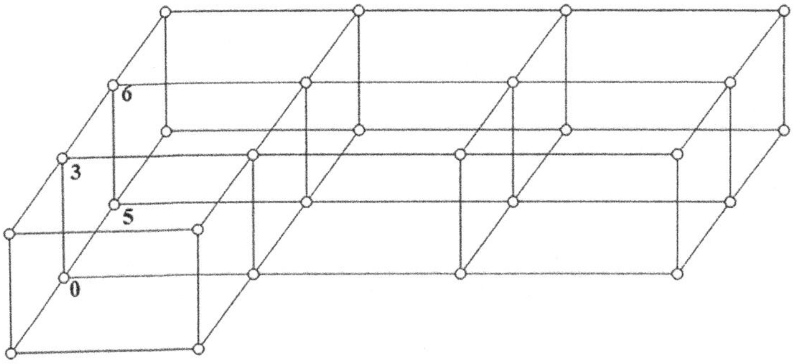

Puzzle 33: Base group: $Z_{28}$

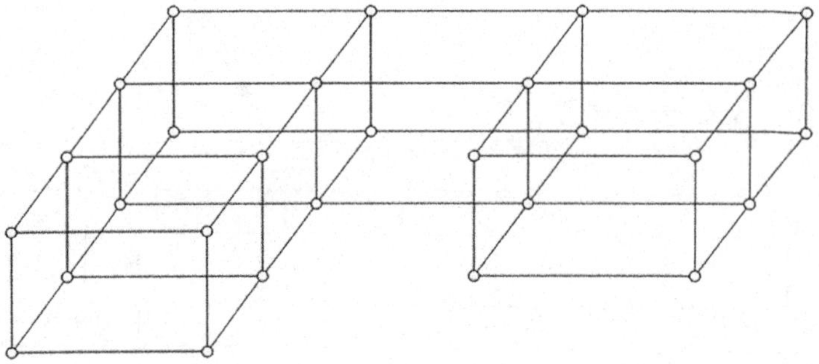

Puzzle 34: Base group $Z_{28}$

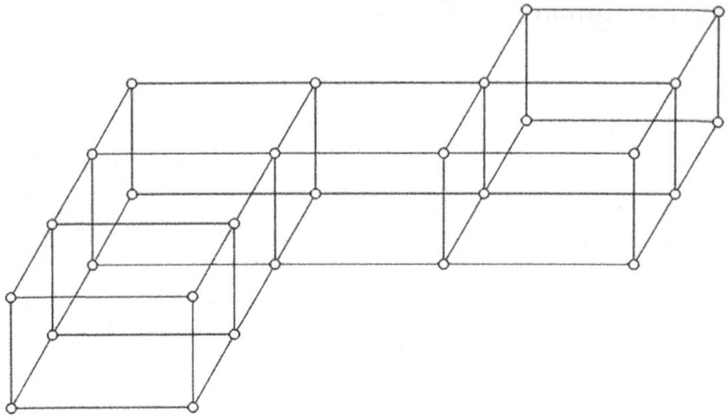

Puzzle 35: Base group $Z_{28}$

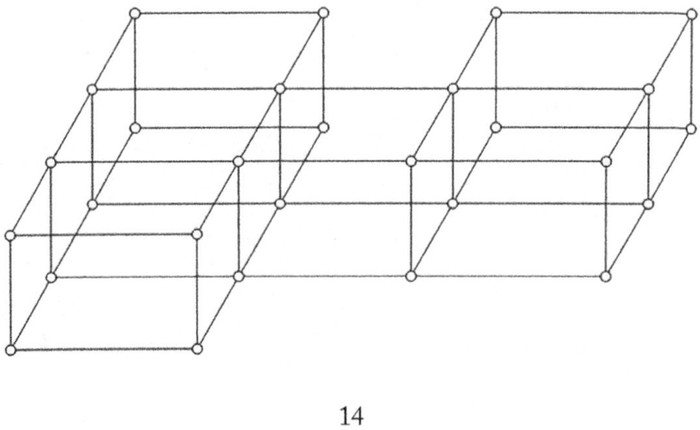

Puzzle 36: Base group $Z_{28}$

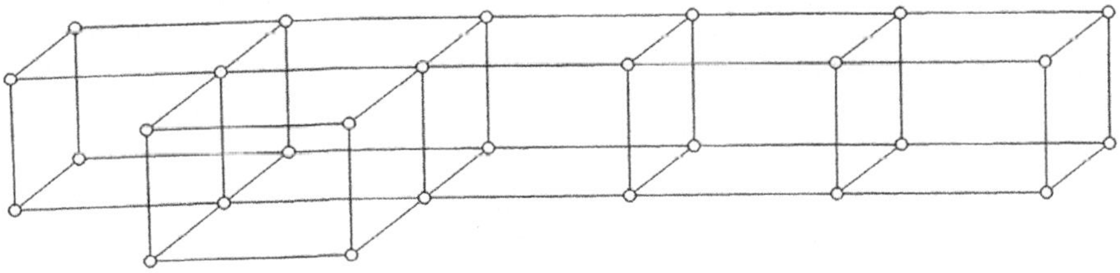

Puzzle 37: Base group $Z_{28}$

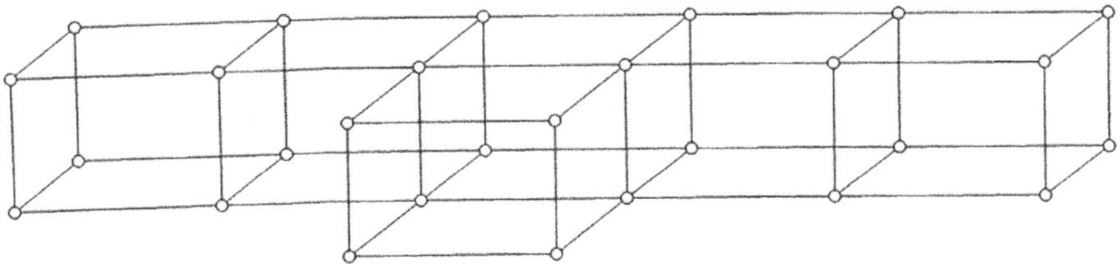

Puzzle 38: Base group $Z_{28}$

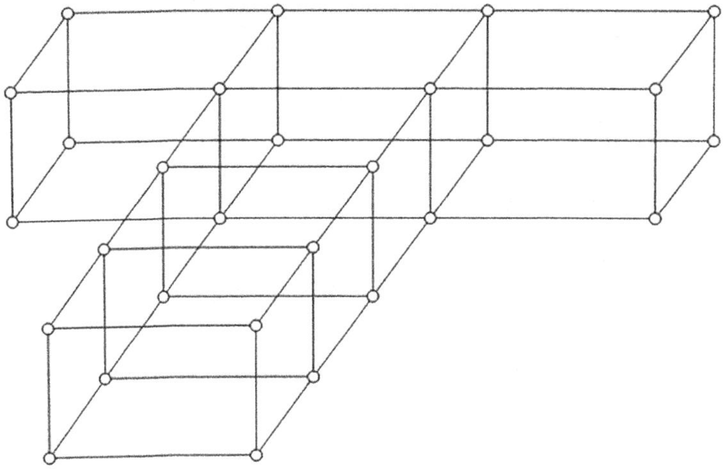

Puzzle 39: Base group $Z_{28}$

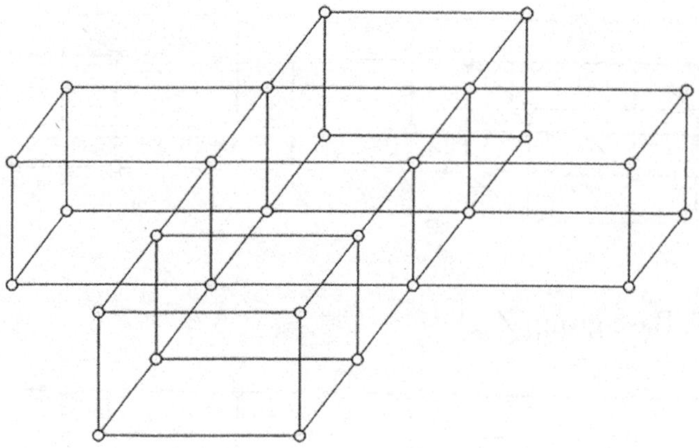

Puzzle 40: Base group $Z_{28}$

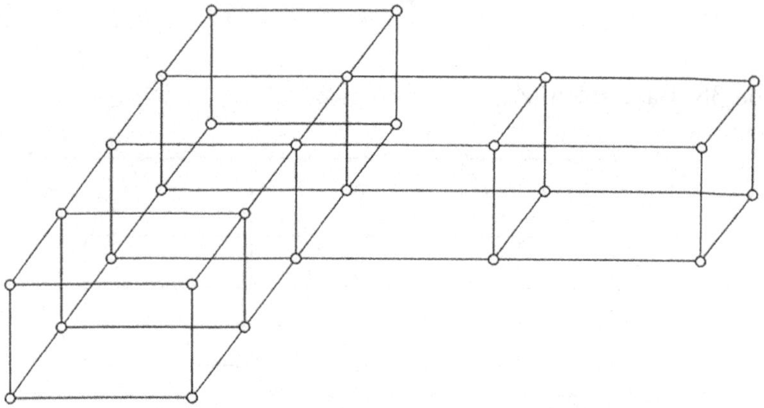

Puzzle 41: Base group $Z_{28}$

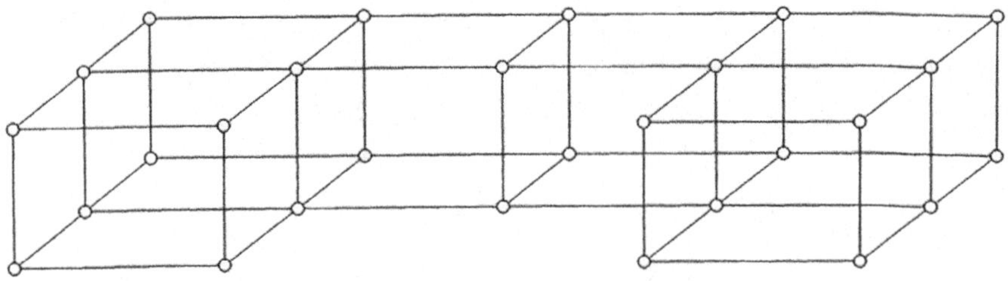

Puzzle 42: Base group $Z_{28}$

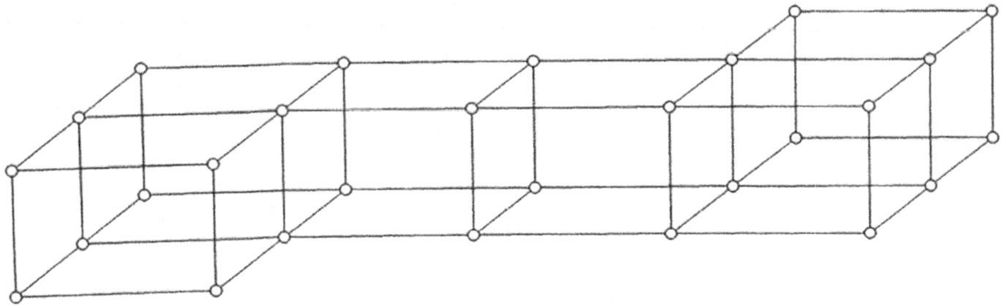

Puzzle 43: Base group $Z_{28}$

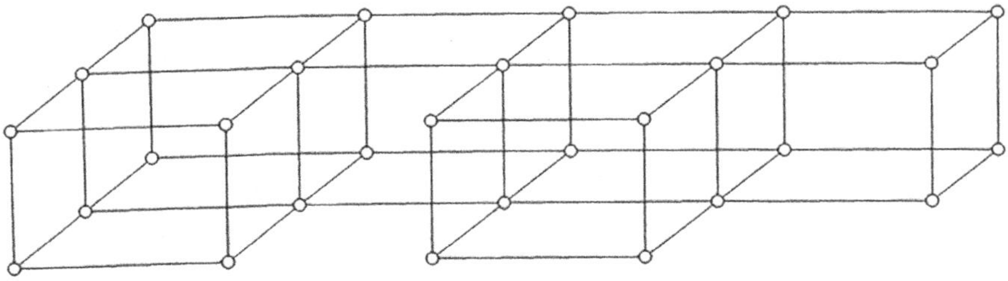

Puzzle 44: Base group $Z_{28}$

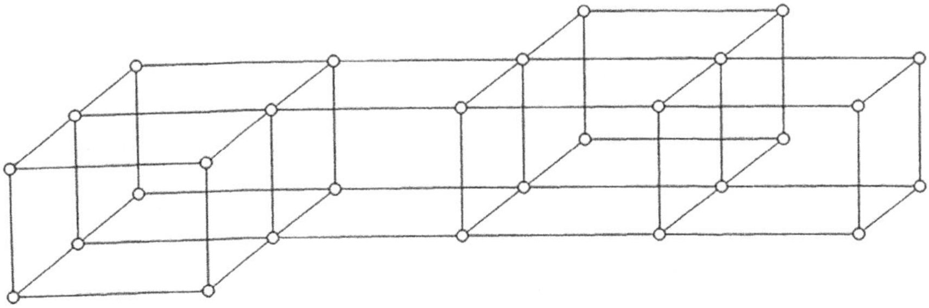

Puzzle 45: Base group $Z_{28}$

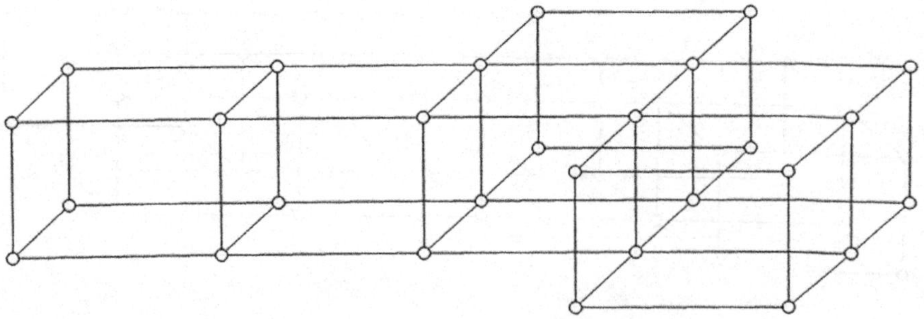

Puzzle 46: Base group $Z_{28}$

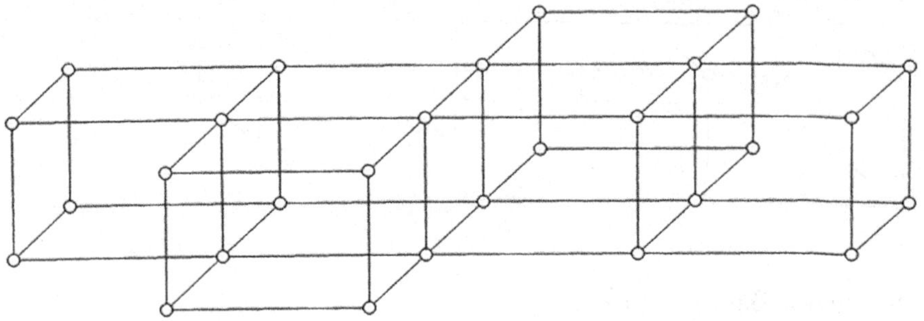

Puzzle 47: Base group $Z_{28}$

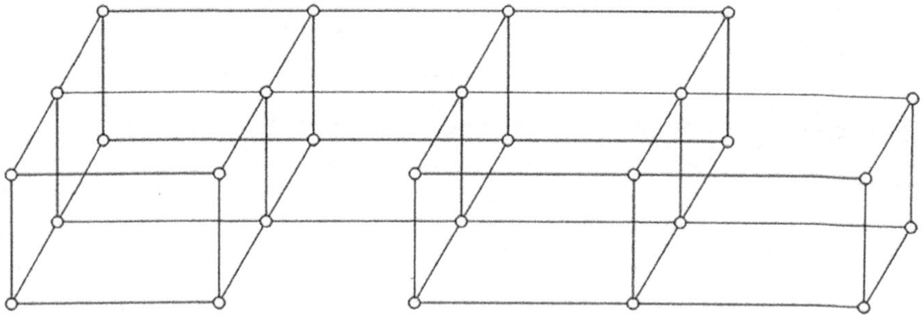

Puzzle 48: Base group $Z_{28}$

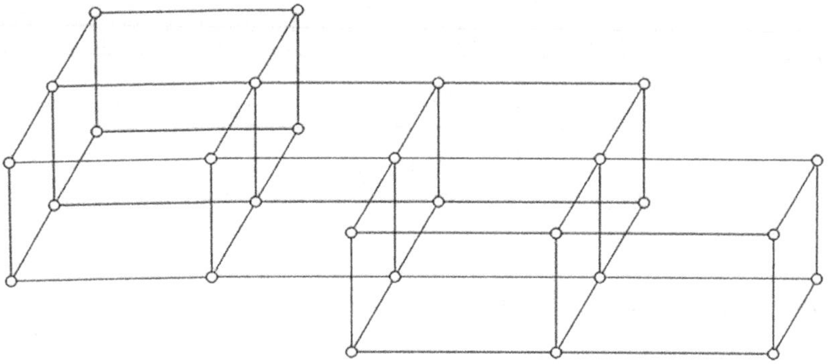

Puzzle 49: Base group $Z_{28}$

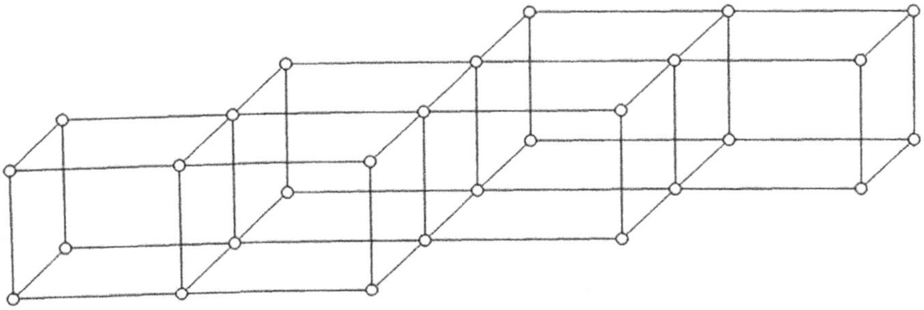

Puzzle 50: Base group $Z_{28}$

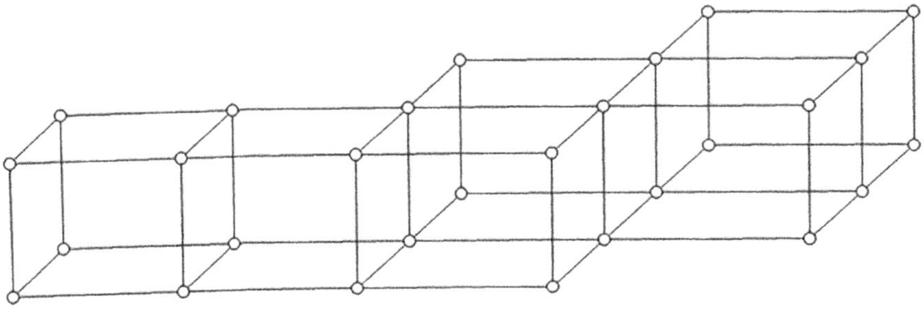

Puzzle 51: Base group $Z_{28}$ ; F = 14

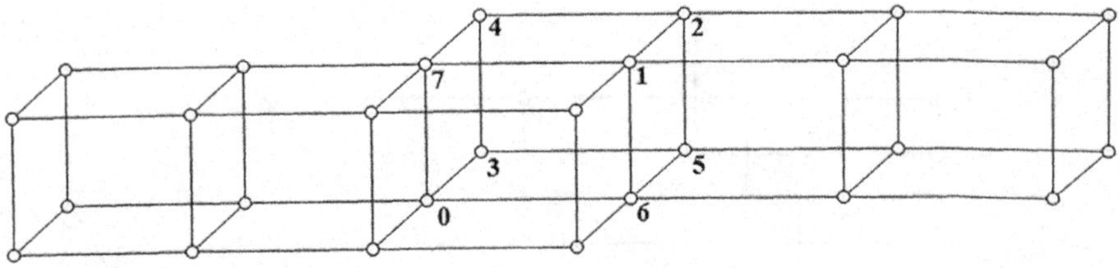


Puzzle 52: Base group $Z_{28}$ ; F = 14

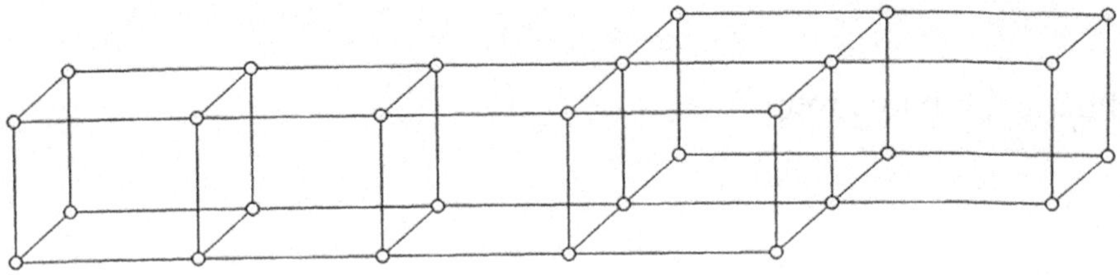

Puzzle 53: Base group $Z_{28}$

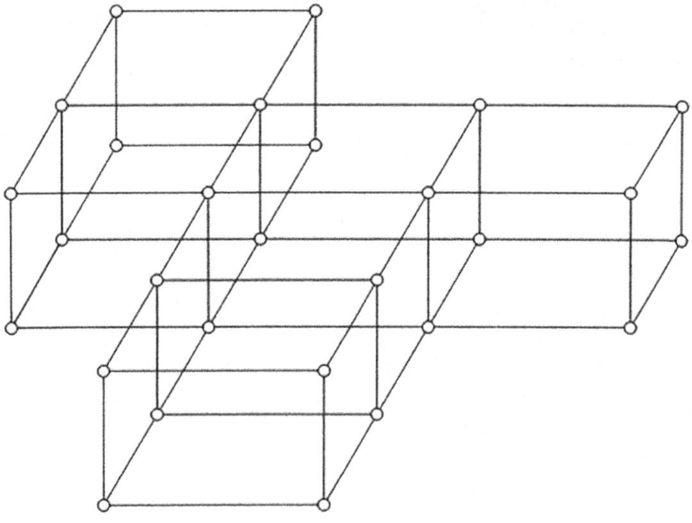

Puzzle 54: Base group $Z_{28}$

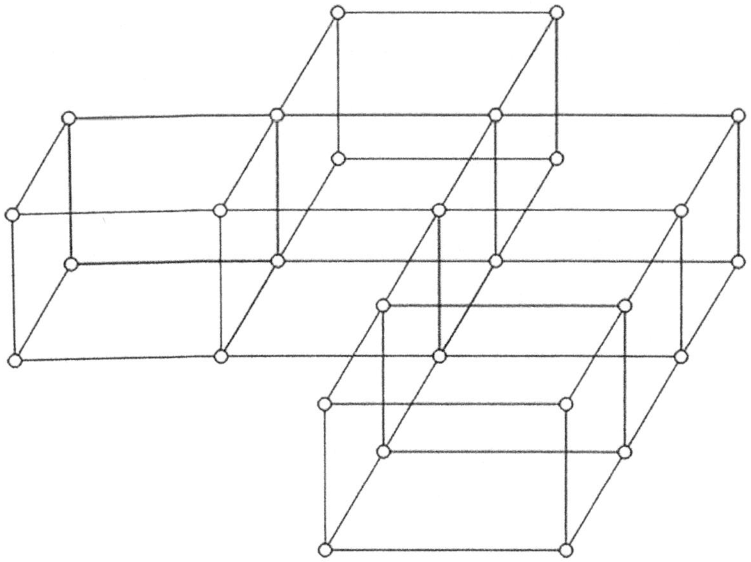

Puzzle 55: Base group $Z_{28}$

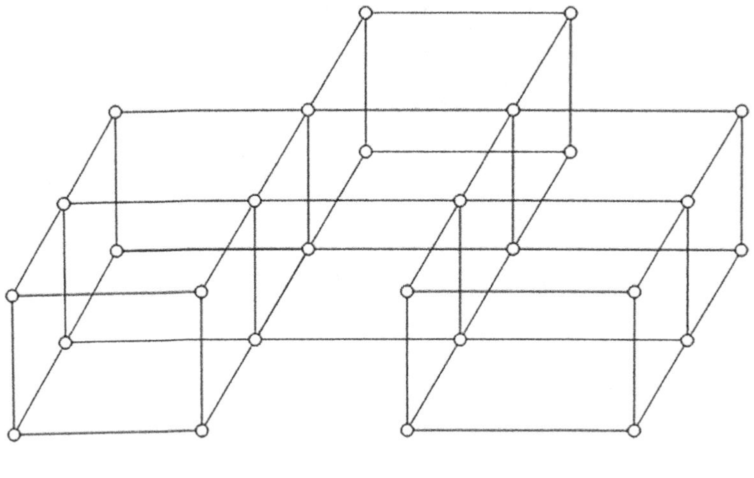

Puzzle 56: Base group $Z_{28}$

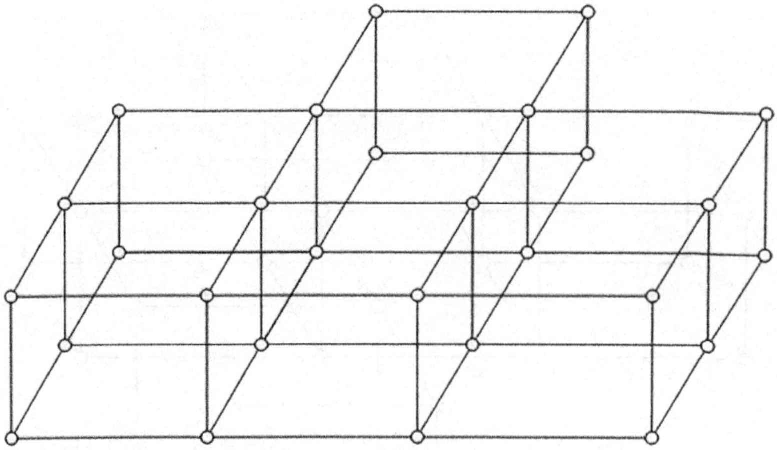

Puzzle 57: Base group $Z_{28}$

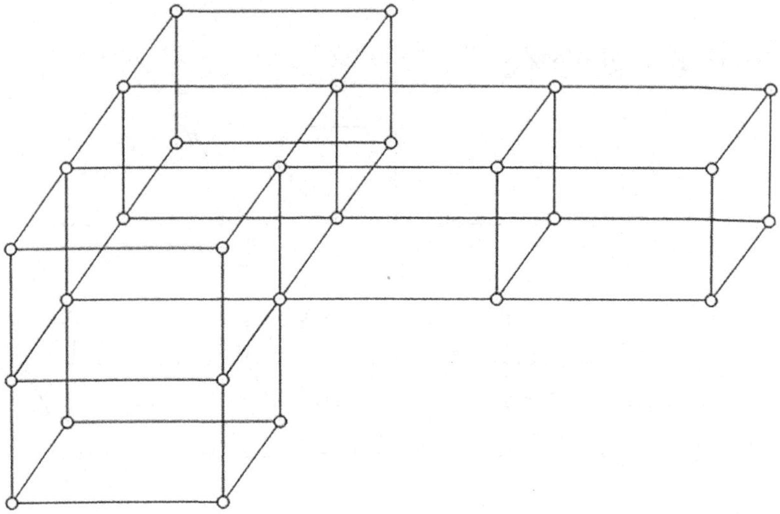

Puzzle 58: Base group $Z_{28}$

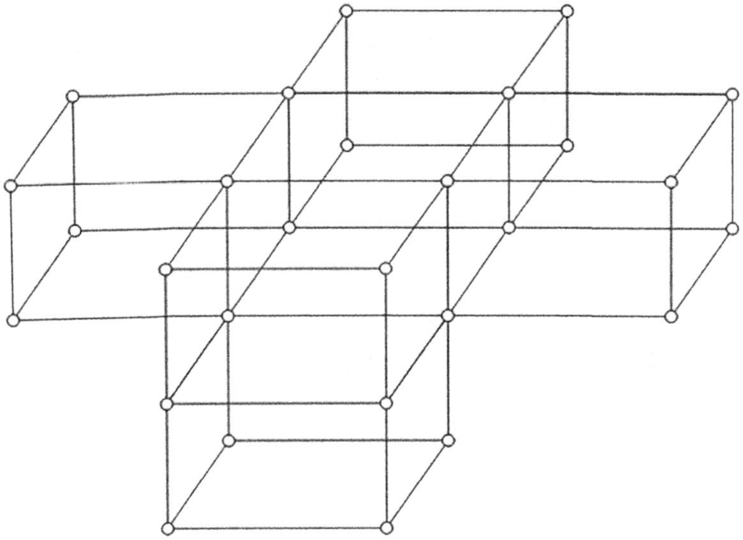

Puzzle 59: Base group $Z_{28}$

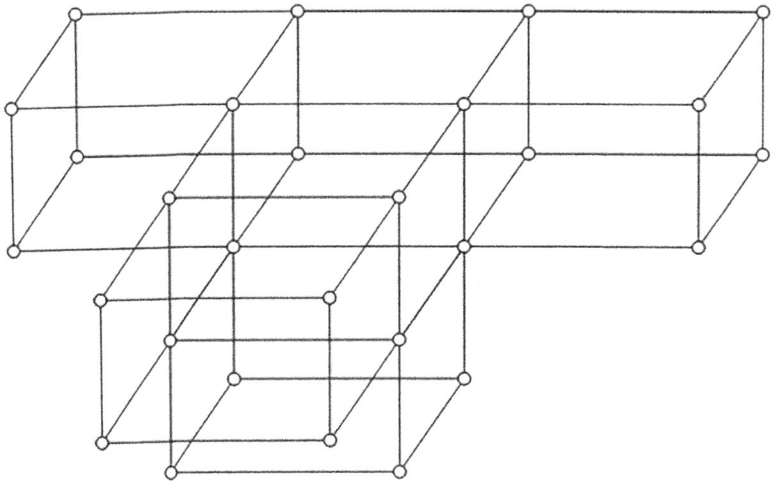

Puzzle 60: Base group $Z_{28}$

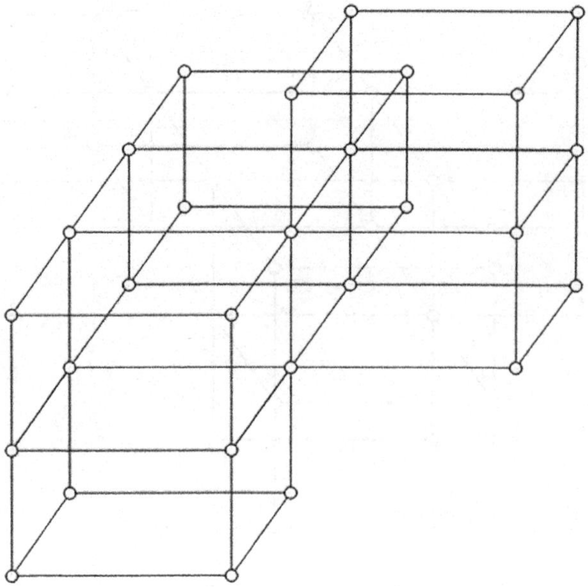

Puzzle 61: Base group $Z_{28}$

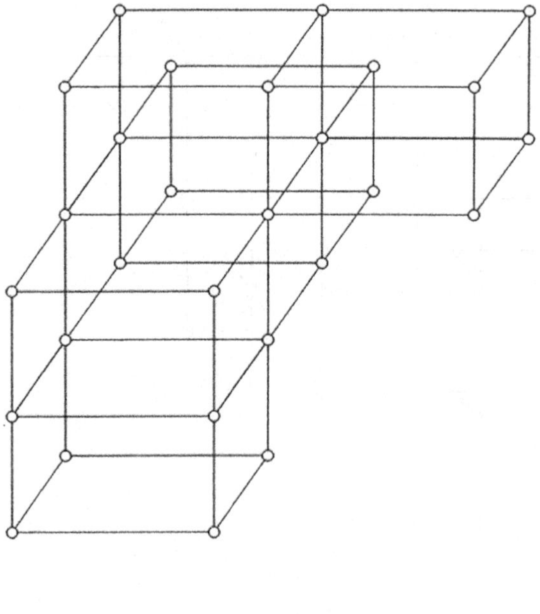

Puzzle 62: Base group $Z_{28}$

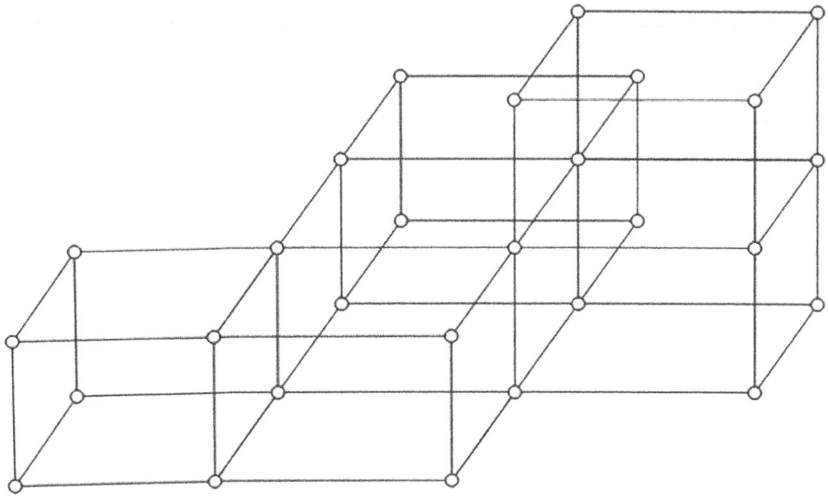

Puzzle 63: Base group $Z_{28}$

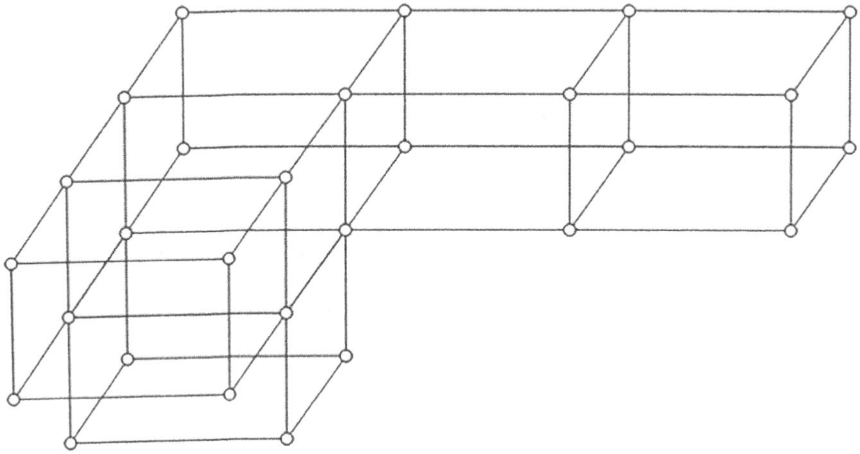

Puzzle 64: Base group $Z_{28}$

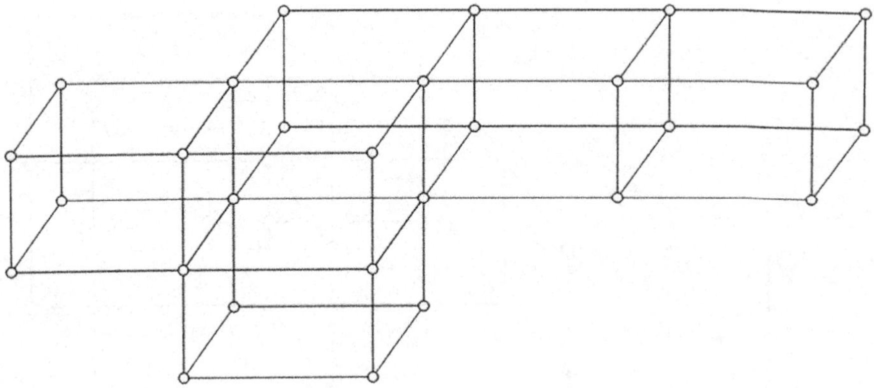

Puzzle 65: Base group $Z_{28}$

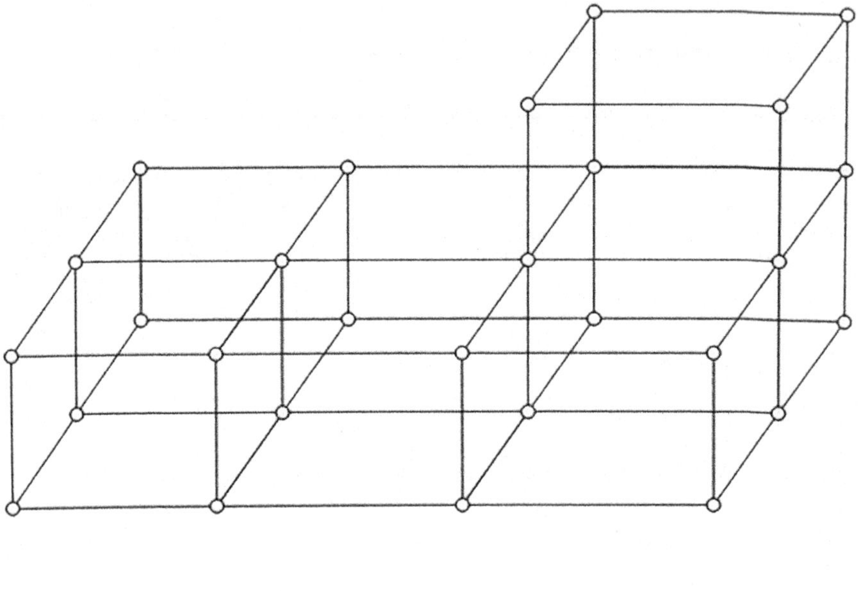

Puzzle 66: Base group $Z_{28}$

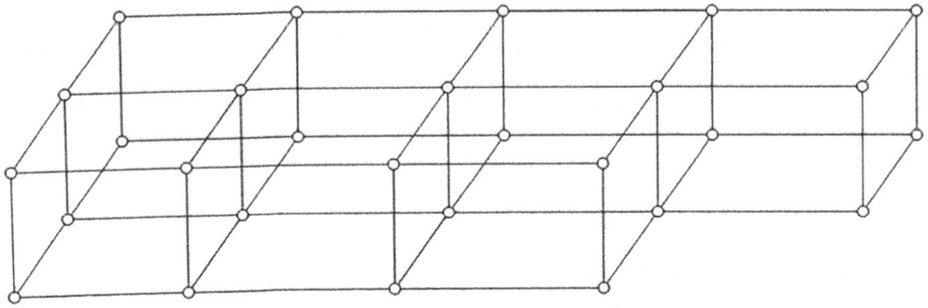

Puzzle 67: Base group $Z_{28}$

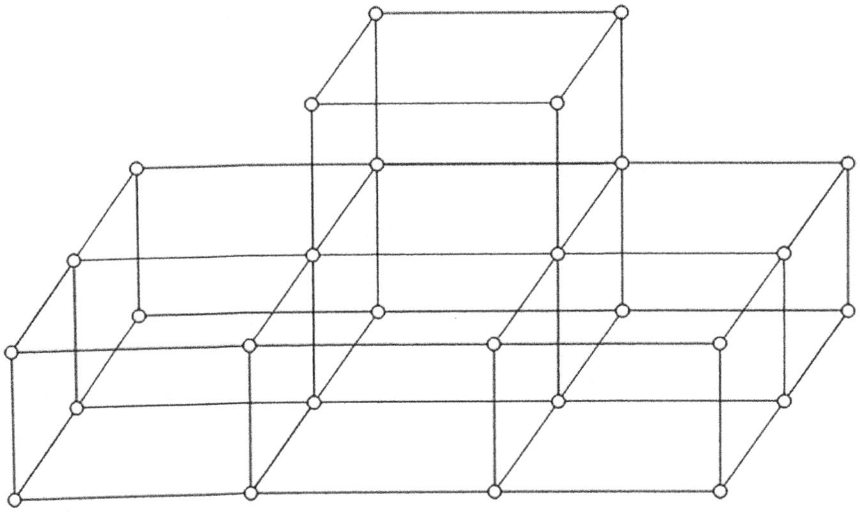

Puzzle 68: Base group $Z_{28}$

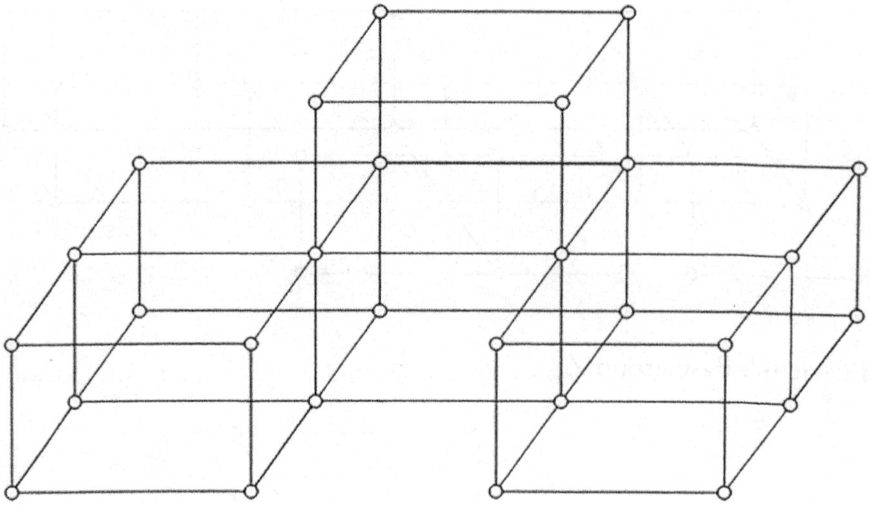

Puzzle 69: Base group $Z_{28}$

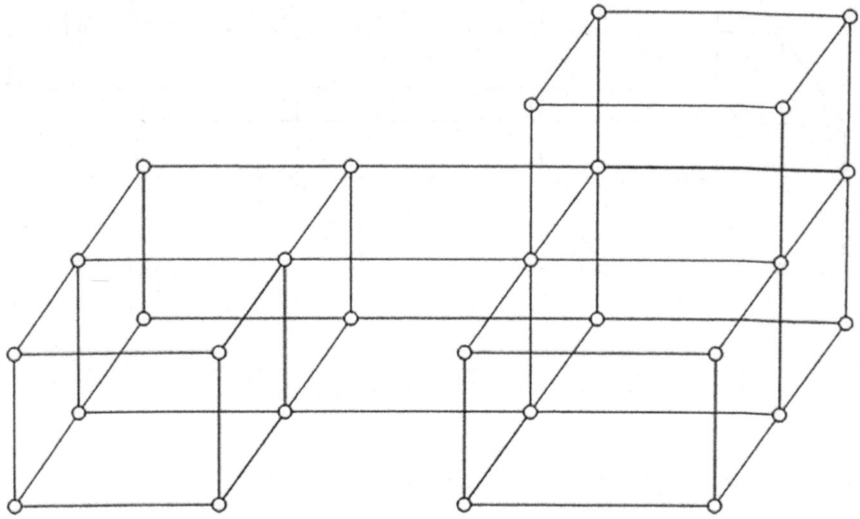

Puzzle 70: Base group $Z_{28}$

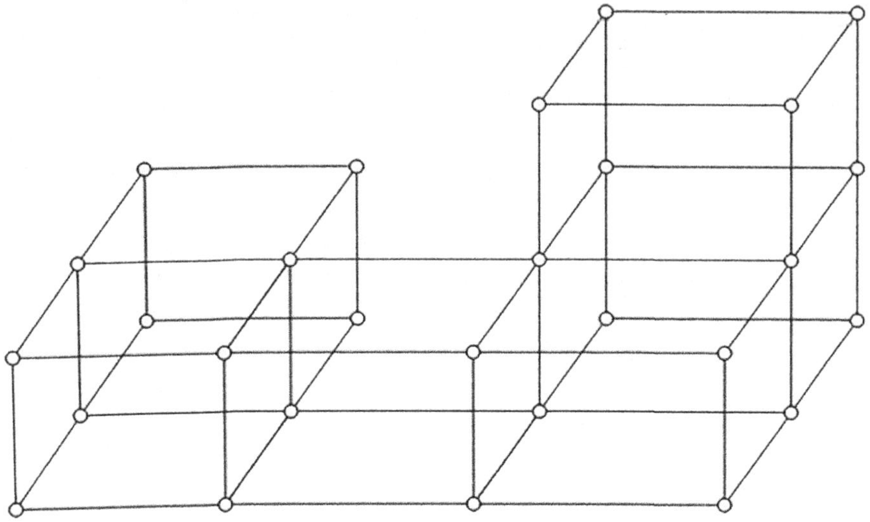

# 5. Rectangular Polycubes

Puzzle 71: Base group $Z_{18}$

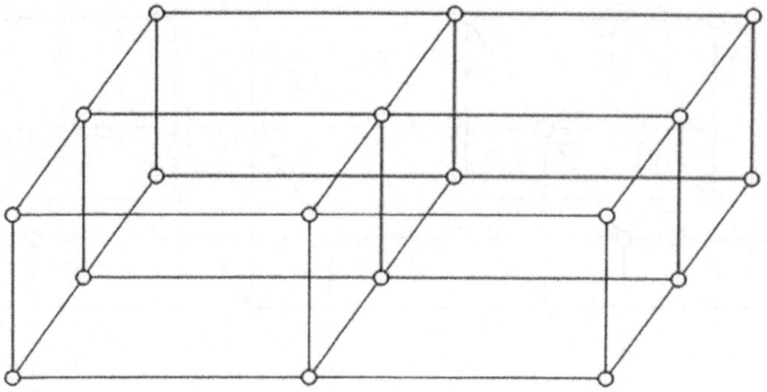

Puzzle 72: Base group $Z_{24}$

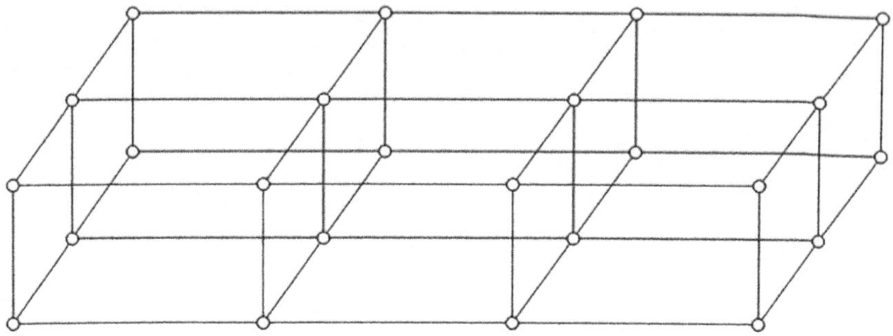

Puzzle 73: Base group $Z_{30}$

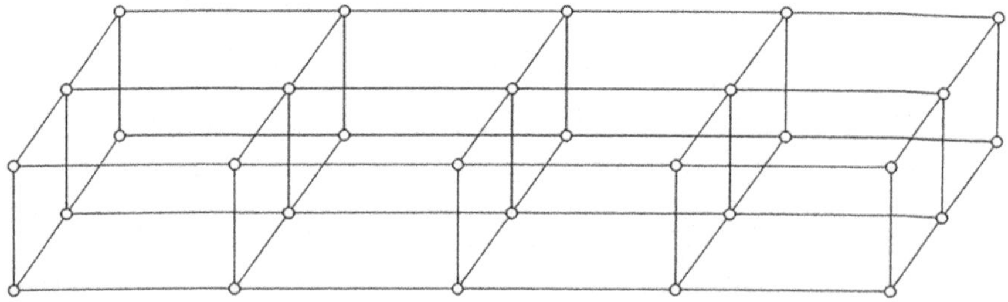

Puzzle 74: Base group $Z_{32}$

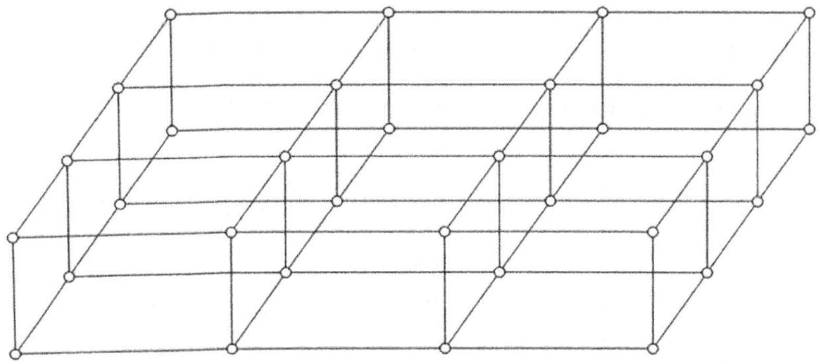

Puzzle75: Base group $Z_{40}$

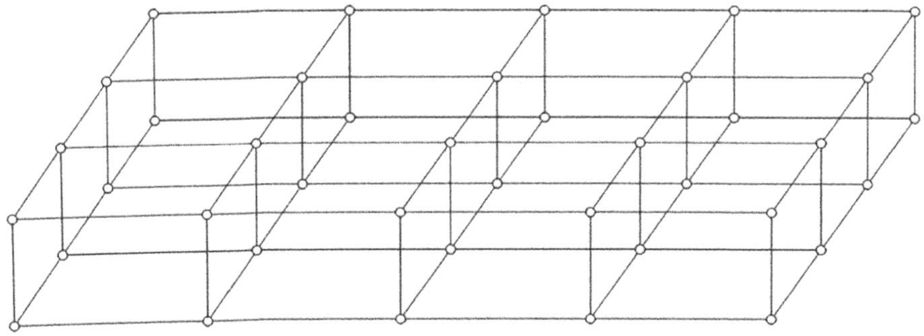

Puzzle 76: Base group $Z_{50}$

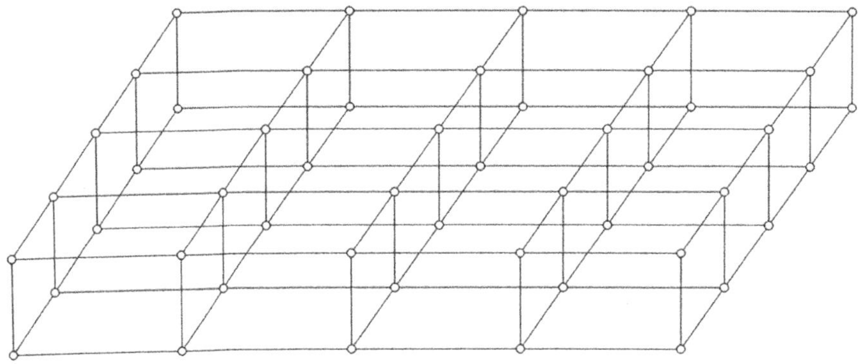

# 6. Box Polycubes

Puzzle 77: Base group $Z_{36}$

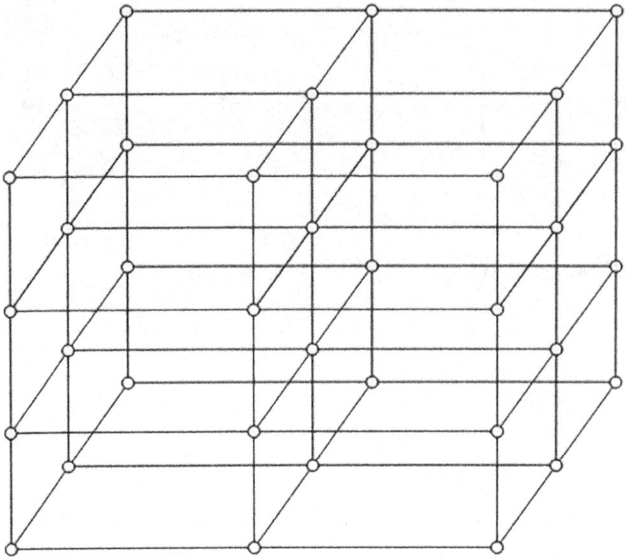

Puzzle 78: Base group $Z_{48}$

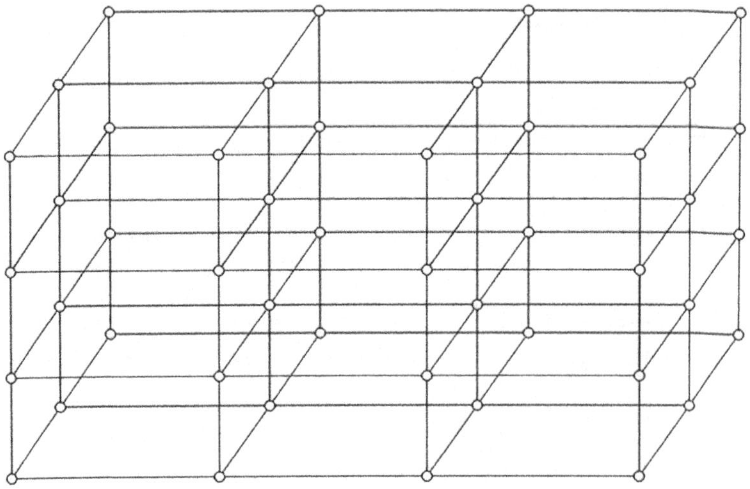

Puzzle 79: Base group $Z_{64}$

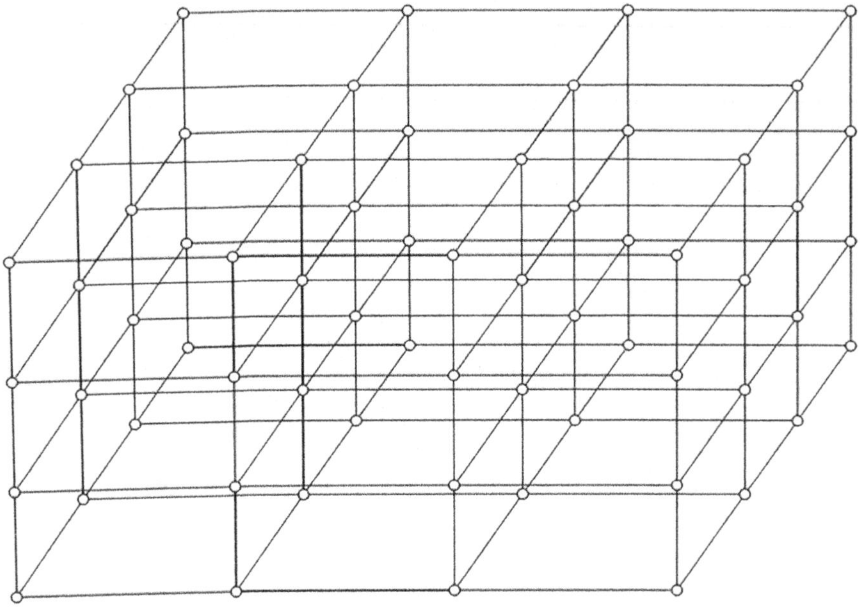

# 7. Staircases

(See also puzzles 9, 21, 49.)

Puzzle 80: Base group $Z_{28}$ ; F = 14

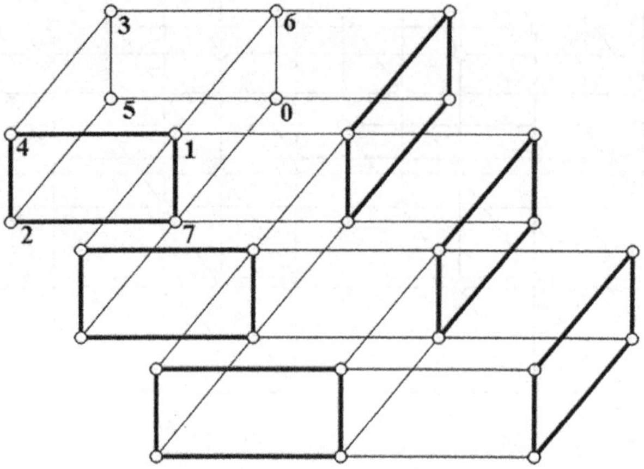


Puzzle 81: Base group $Z_{36}$ ; F = 14

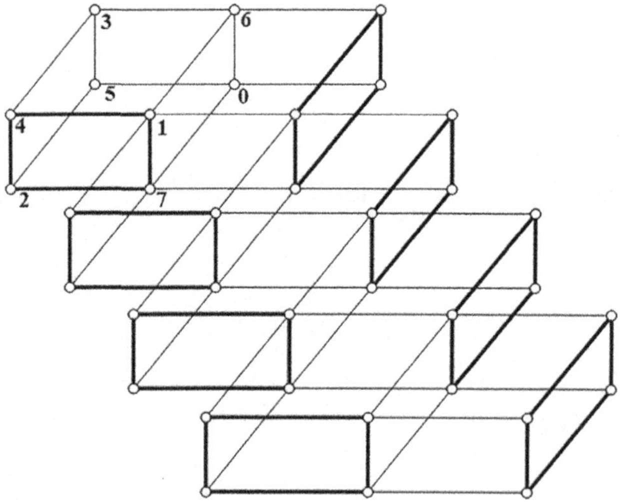

Puzzle 82: Base group $Z_{44}$ ; F = 14

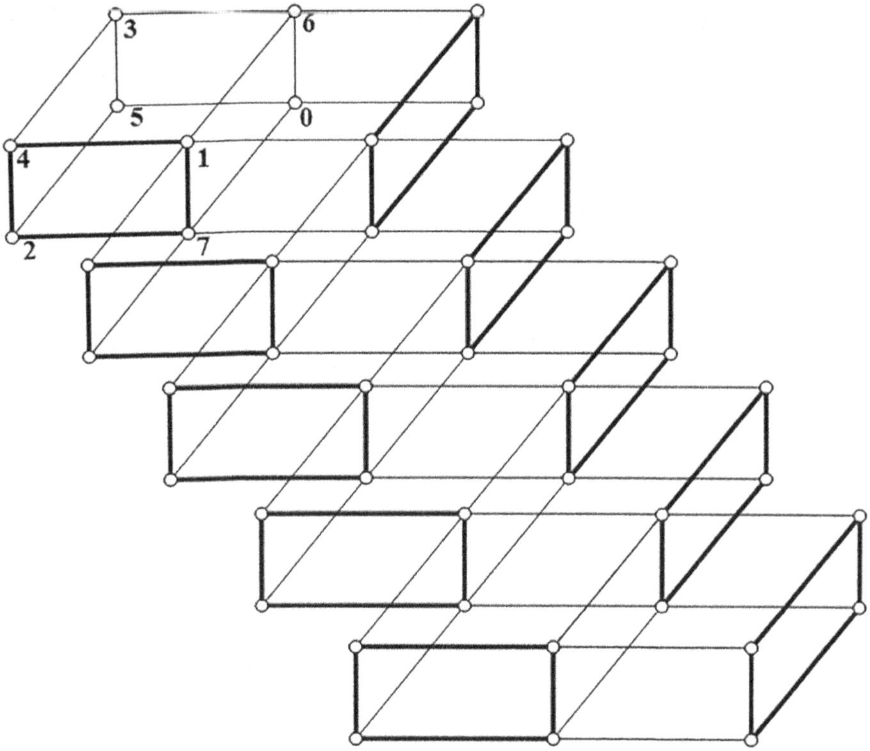

Puzzle 83: Base group $Z_{28}$ ; F = 14

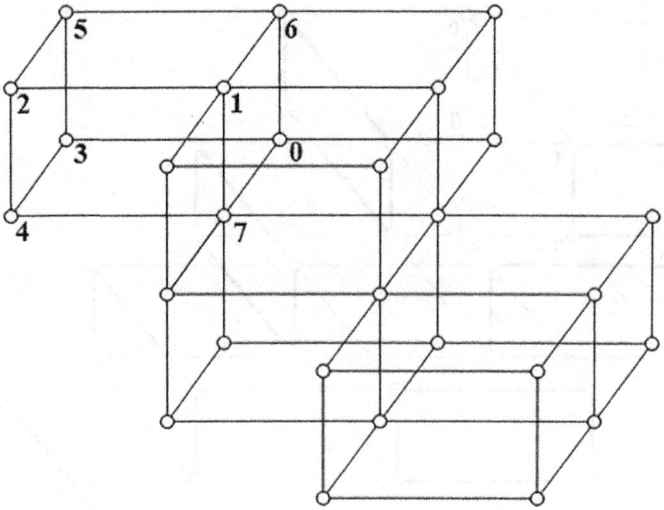


Puzzle 84: Base group $Z_{32}$ ; F = 14

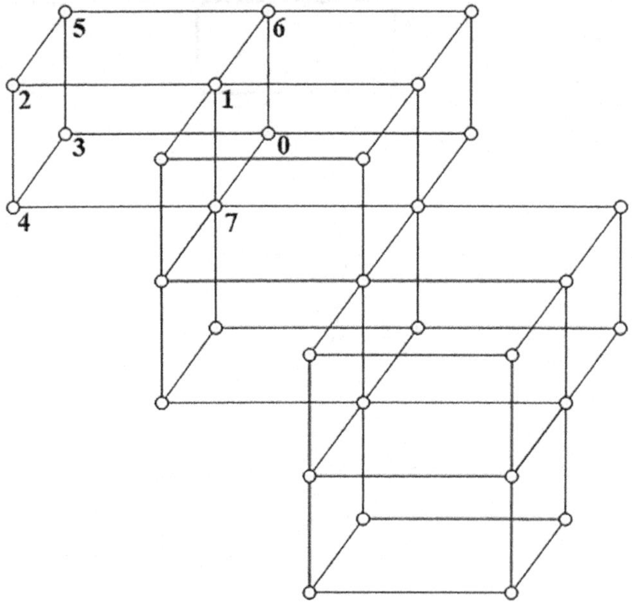

Puzzle 85: Base group $Z_{36}$ ; F = 14

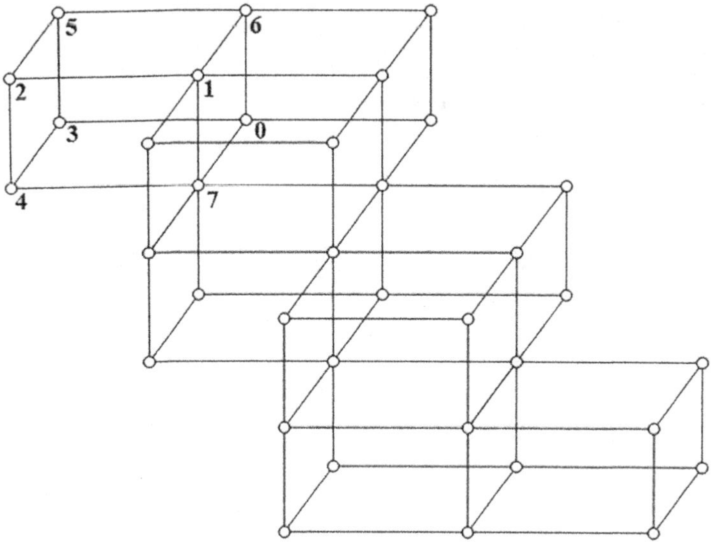


Puzzle 86: Base group $Z_{40}$ ; F = 14

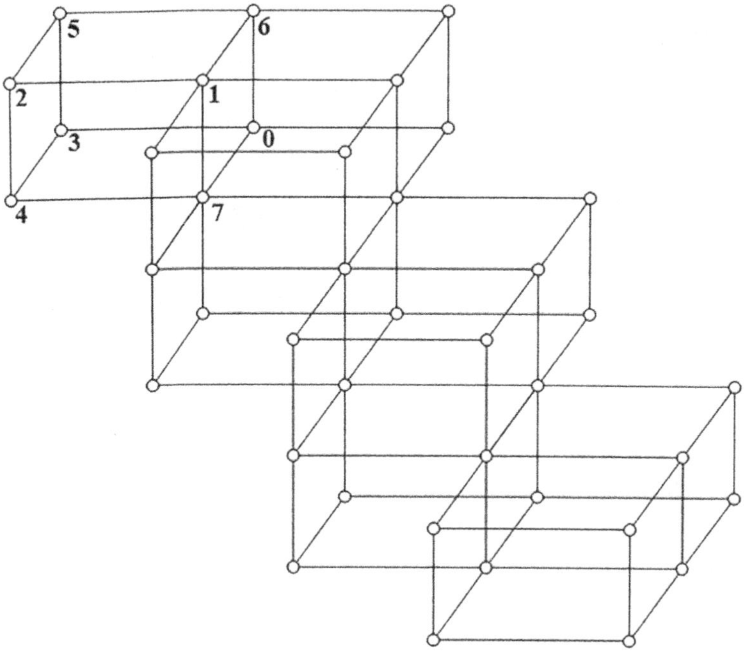

# 8. Tripods and Stars

Puzzle 87: Base group $Z_{20}$ ; F = 14

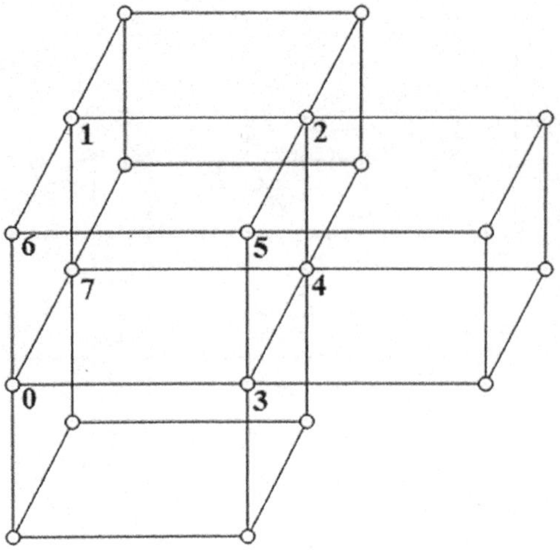


Puzzle 88: Base group $Z_{24}$ ; F = 14

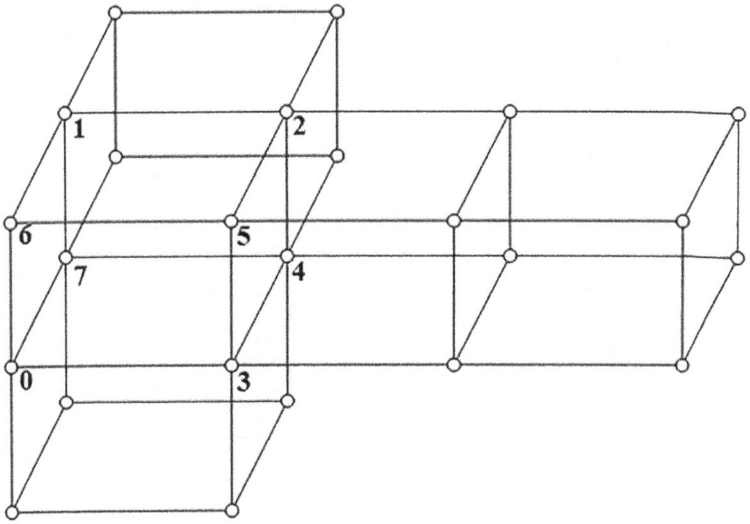

Puzzle 89: Base group $Z_{28}$ ; F = 14

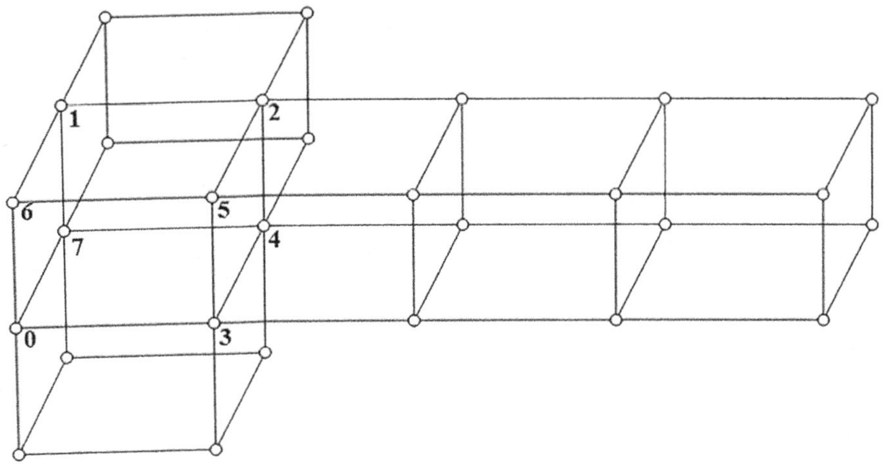


Puzzle 90: Base group $Z_{28}$ ; F = 14

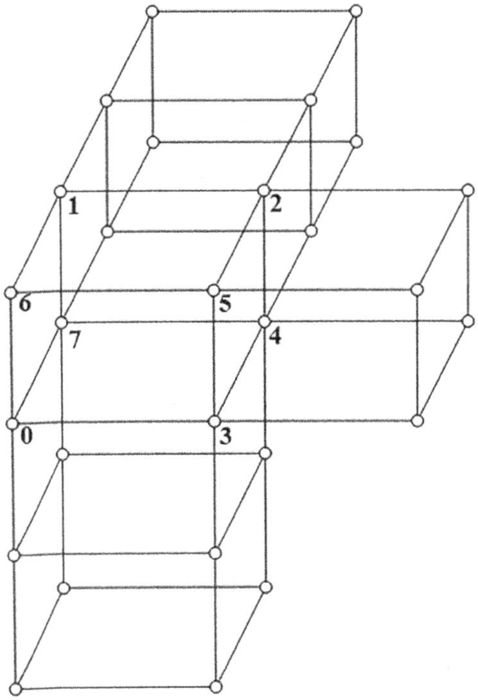

Puzzle 91: Base group $Z_{32}$ ; F = 14

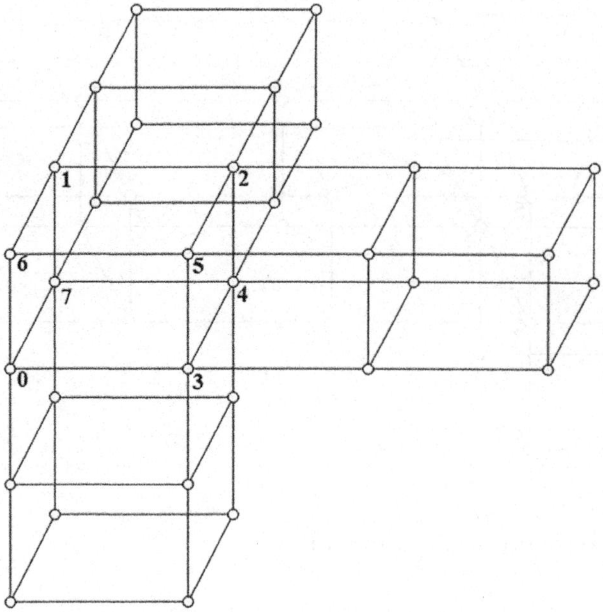


Puzzle 92: Base group $Z_{32}$ ; F = 14

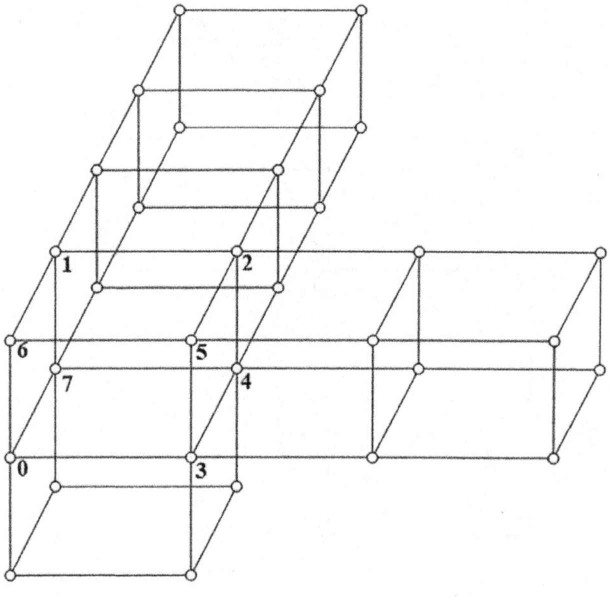

Puzzle 93: Base group $Z_{36}$ ; F = 14

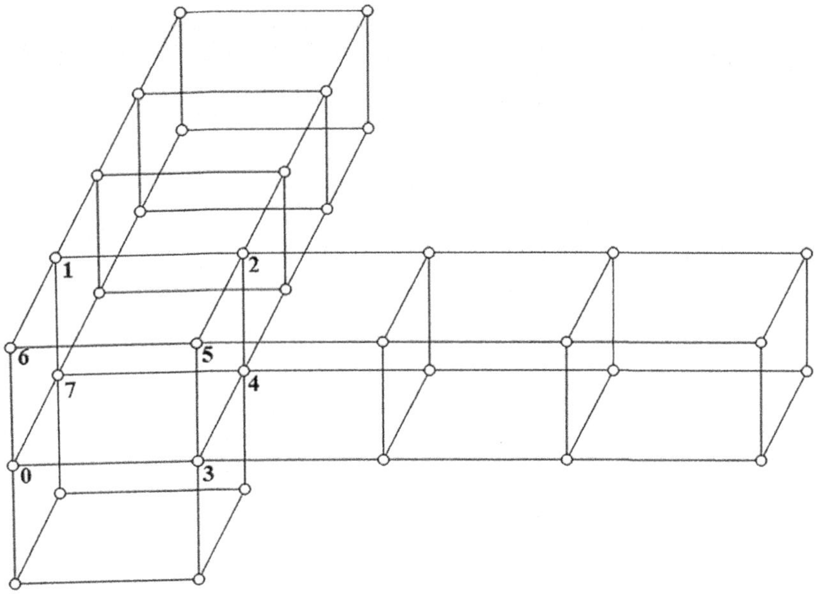


Puzzle 94: Base group $Z_{36}$ ; F = 14

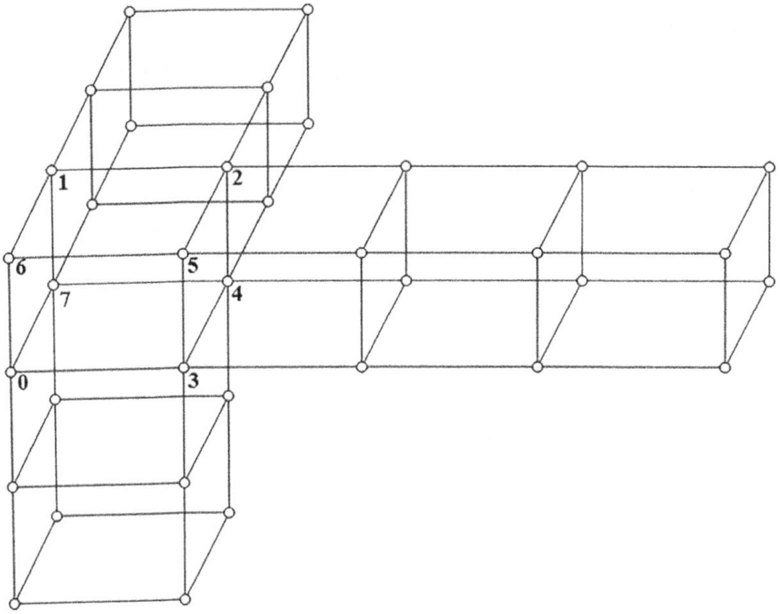

Puzzle 95: Base group $Z_{24}$ ; F = 14

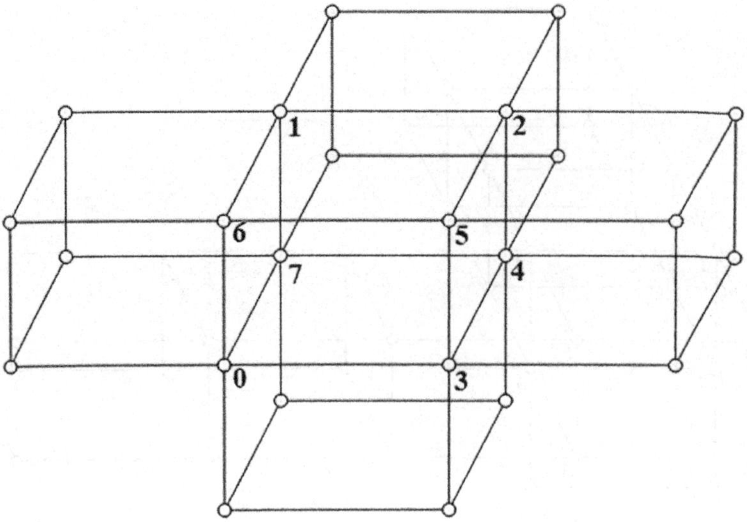


Puzzle 96: Base group $Z_{28}$ ; F = 14

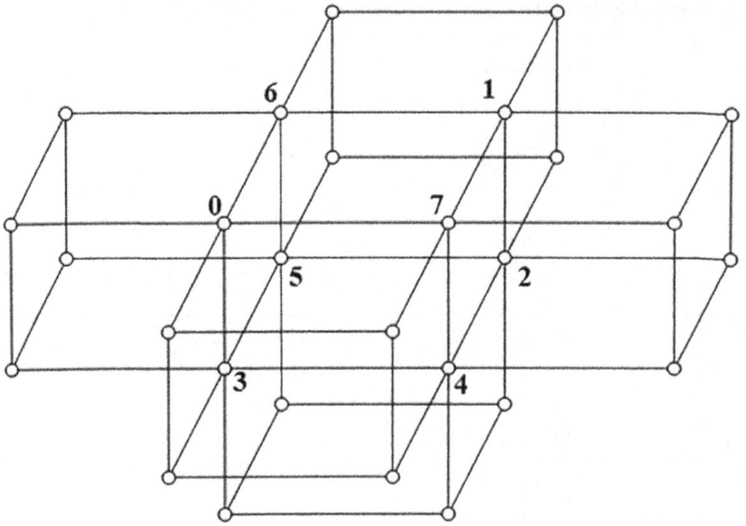

Puzzle 97: Base group $Z_{32}$ ; F = 14

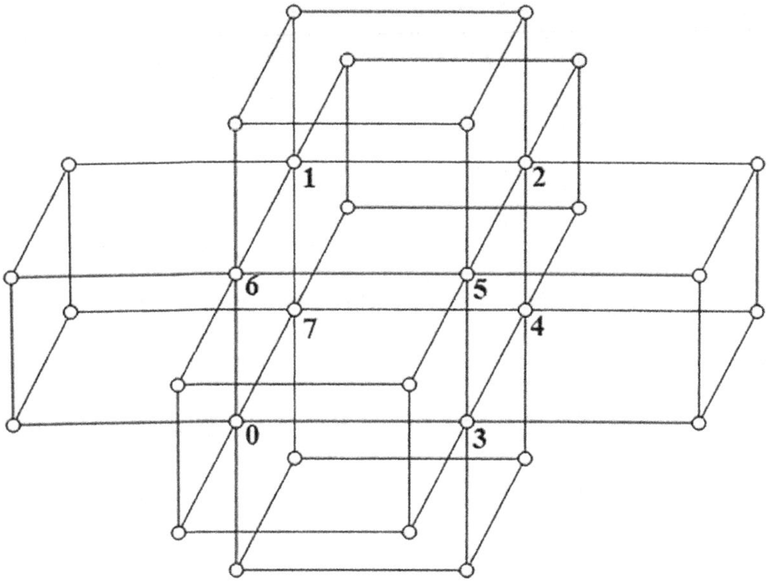


Puzzle 98: Base group $Z_{36}$ ; F = 14

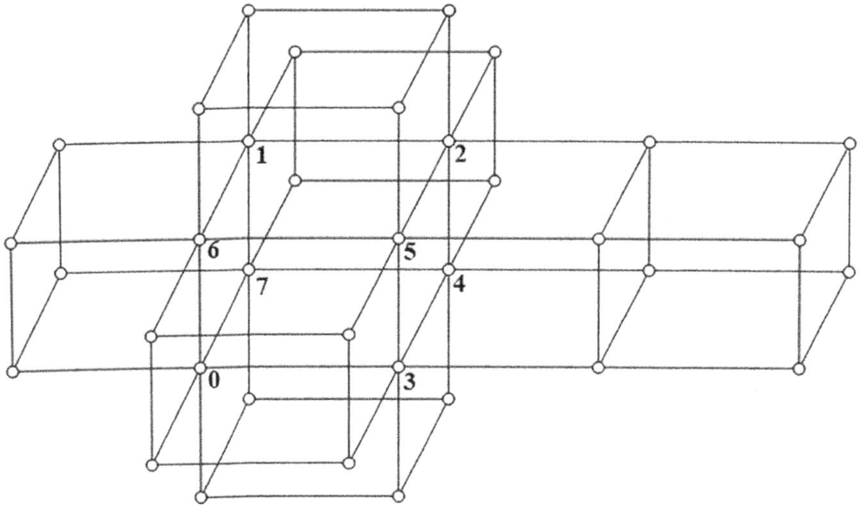

Puzzle 99: Base group $Z_{44}$ ; F = 14

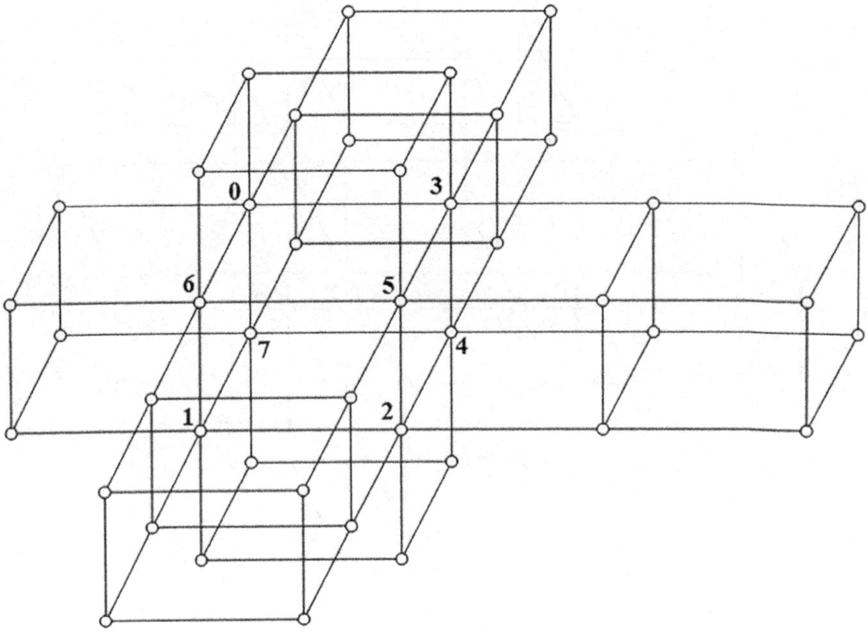

# 9. String Polycubes

Puzzle 100: Base group $Z_{80}$ = {0, 1,-1,2,-2, …, 39,-39, 40}

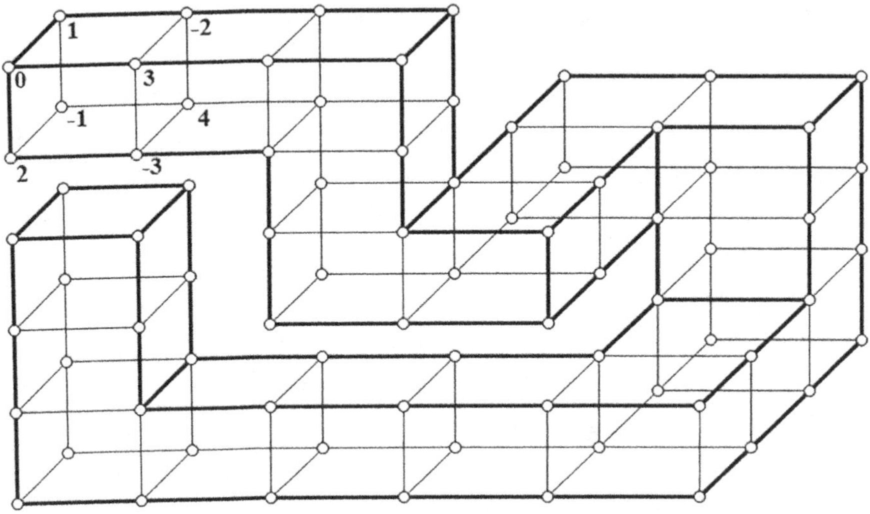

# Polycubes: problems and research questions

1. Show that every linear polycube with n points has a solution over $Z_n$

2. Three directions can be associated with every polycube. The *x-direction* is hiorizontal; the *z-direction* is vertical, and the *y-direction* is the direction perpendicular to the x,y plane. If P is a polycube, then a *linear component* of P is a linear polycube embedded in P. Linear components can occur in the x, y or z-direction. The *length* of a linear component is the number of points of the component which lie in a straight line. Note that the length is one more than the number of cubes in a linear polycube. For example, the polycube of puzzle 99 has three linear components, one in each direction. The one in the x-direction has length 5; the one in the z-direction has length 4, and the one in the y-direction has length 6. A *tree polycube* is a polycube in which no edge belongs to 4 cubes. In Chapters 1,2,3 and 4 every polycube except for the ones in puzzles 32, 56, 65, 66, 67 is a tree polycube. Show that every tree polycube with n points and with even length linear components in the y and z-directions, has a solution over $Z_n$.

3. * Find a method for solving every tree polycube over $Z_n$.

4. * No solution has been found for the following small polycubes. Find solutions or show that no solution exists.

(i) Solve over $Z_{22}$

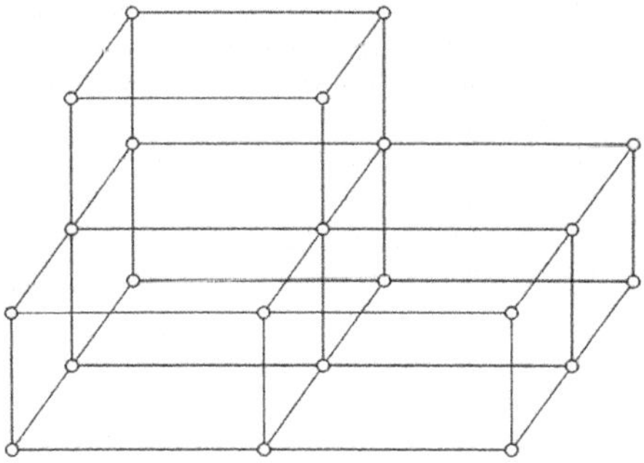

(ii) Solve over $Z_{24}$

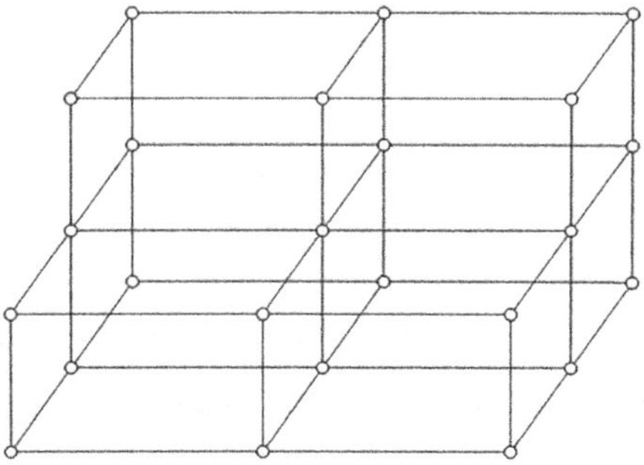

(iii) Solve over $Z_{26}$

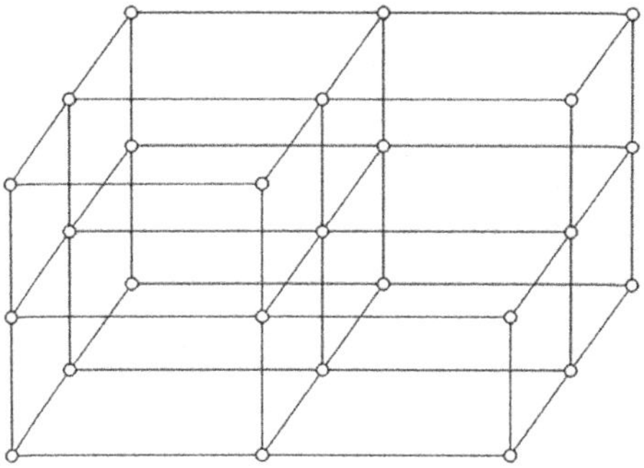

(iv) Solve over $Z_{27}$

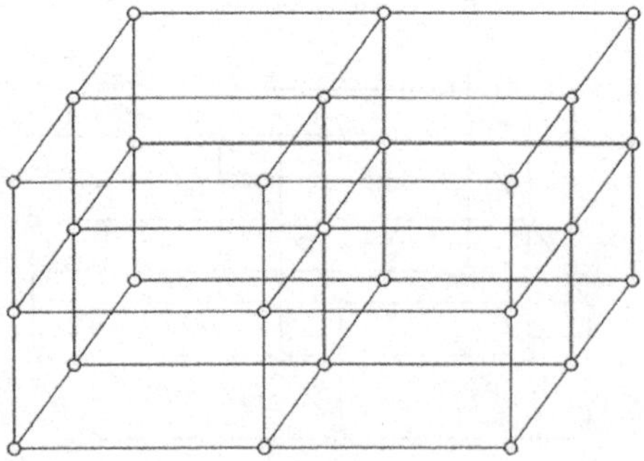

5. A *tripod* is a polycube with three linear components in each of the x, y and z directions, and which share a common endcube. A *star* is a polycube also with three linear components, one in each of the three directions, and which share a common cube, (which need not be an endcube). See Chapter 8. Show that every tripod and every star with n points is solvable over $Z_n$.

6. Puzzles 80,81,82 feature *staircase polycubes* with *stepsize 3.* Show that every such polycube with n points is solvable over $Z_n$. Puzzles 83,84,85,86 feature *skew staircases* with step size 3. Show that every such stacircase with n points is solvable over $Z_n$. See Chapter 7.

7. Show that every planar rectangular polycube with n points has a solution over $Z_n$. See Chapter 5.

8. Show that every box polycube with an even number n of points (hence at least one even dimension), is solvable over $Z_n$. See Chapter 6.

9. * No solutions have been found for any box with an odd number of points, (hence three odd dimensions). Find solutions or show that no solution exists.

10. A *string polycube* is a polycube in which each component cube, except two *endcubes,* shares a face with exactly two other cubes. Note that in the string polycube (puzzle 100) featured in Chapter 9, every linear component in the y and z directions has even length. (Recall that the *length* of a linear polycube is the number of points in a line on the edge of the polycube.) Show that every string polycube with n points and with even length linear components in the y and z directions is solvable over $Z_n$.

11. * Find a method for solving every string polycube.

12. * A *ring* is a string polycube but without endcubes. Thus every component cube of a ring shares a face with exactly two other cubes. By the addition of four edges the string given in puzzle 100 can be converted to a ring. Find a method, if possible, to solve every ring or at least every ring with even linear components in two directions.

SECTION II

# TRIANGULATIONS

# 10. Triangulations over $Z_6$ and $Z_7$

Puzzle 101: Base group $Z_6$

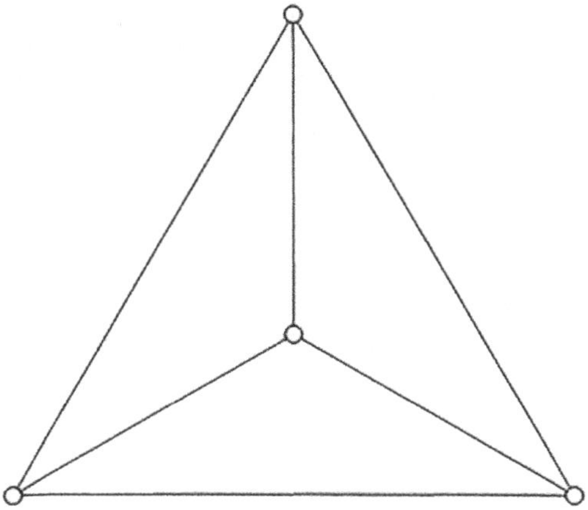

Puzzle 102: Base group $Z_7$

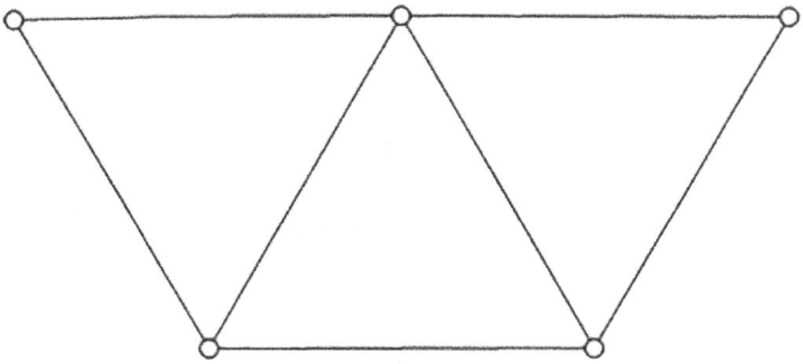

# 11. Triangulations over $Z_8$

Puzzle 103: Base group $Z_8$

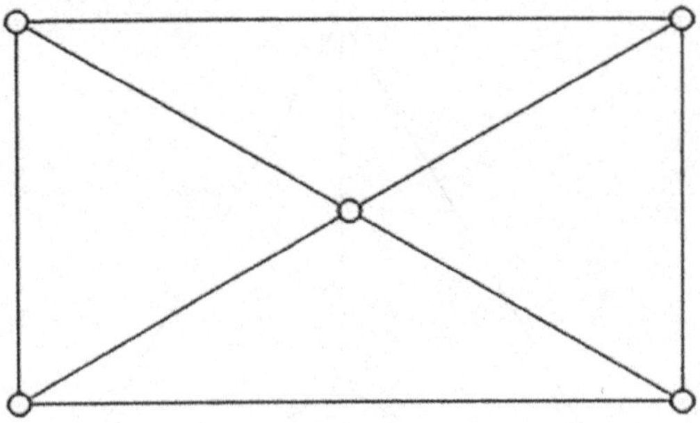

Puzzle 104: Base group $Z_8$

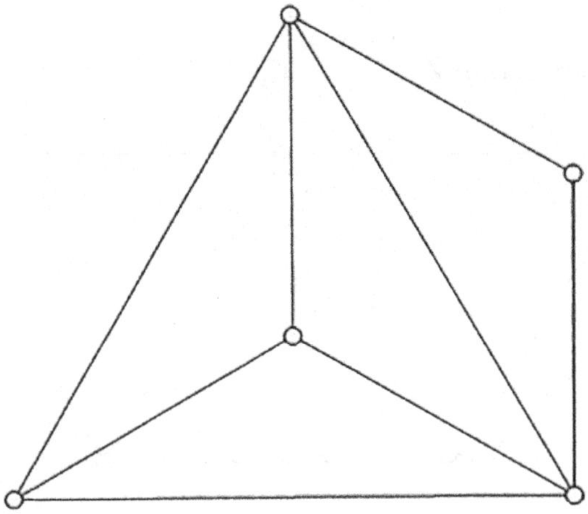

# 12. Triangulations over $Z_9$

Puzzle 105: Base group $Z_9$

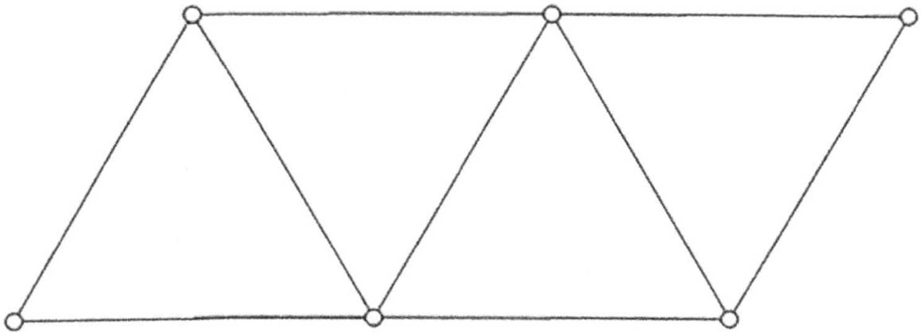

Puzzle 106: Base group $Z_9$

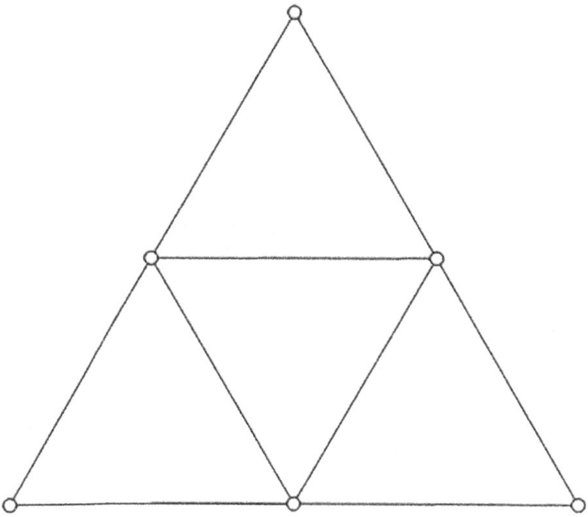

Puzzle 107: Base group $Z_9$

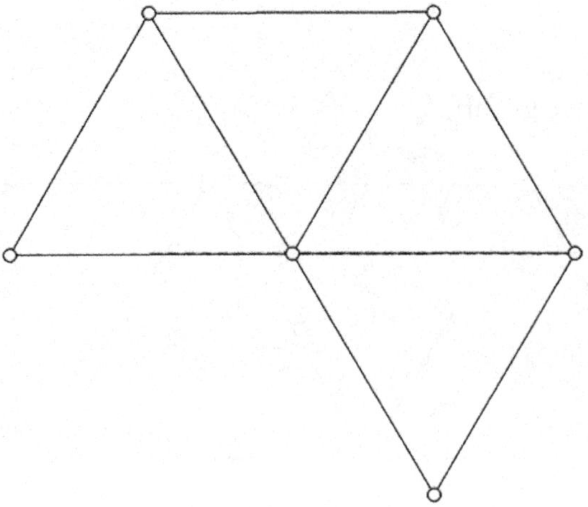

Puzzle 108: Base group $Z_9$

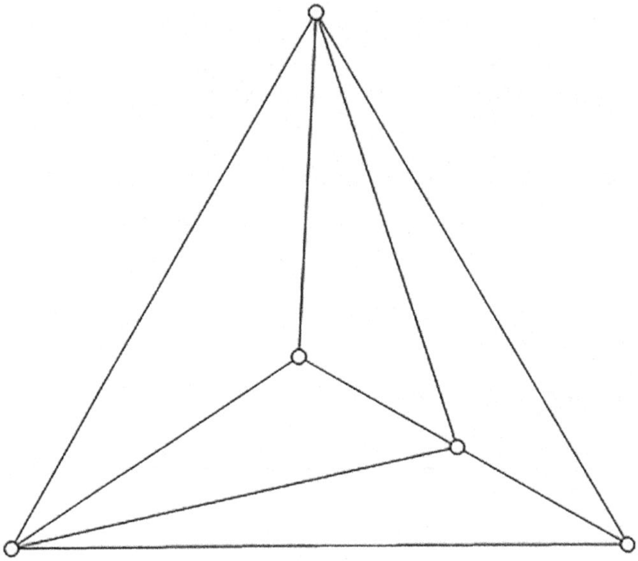

# 13. Triangulations over $Z_{10}$

Puzzle 109: Base group $Z_{10}$

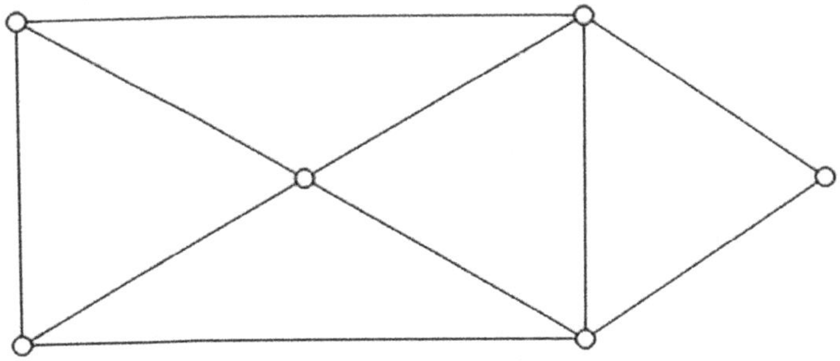

Puzzle 110: Base group $Z_{10}$

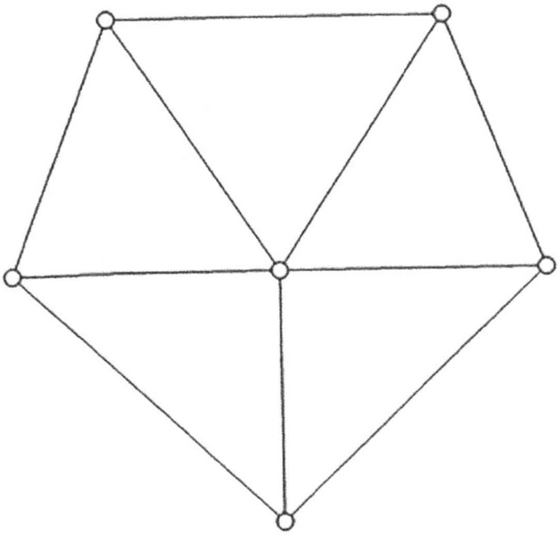

Puzzle 111: Base group $Z_{10}$

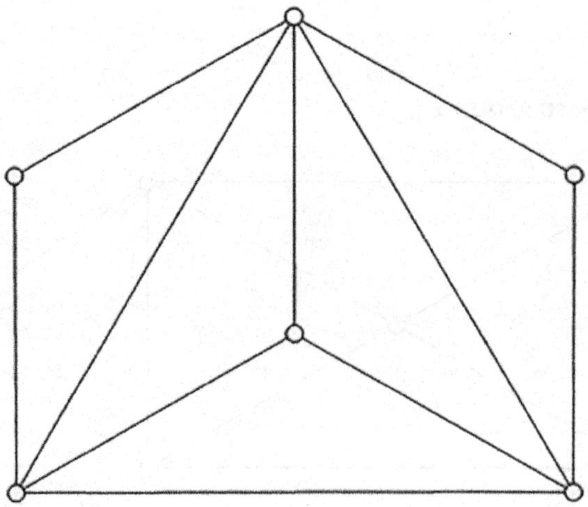

Puzzle 112: Base group $Z_{10}$

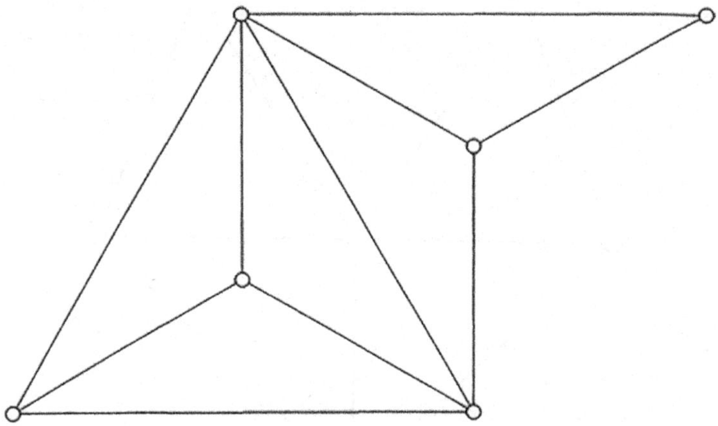

# 14. Triangulations over $Z_{11}$

Puzzle 113: Base group $Z_{11}$

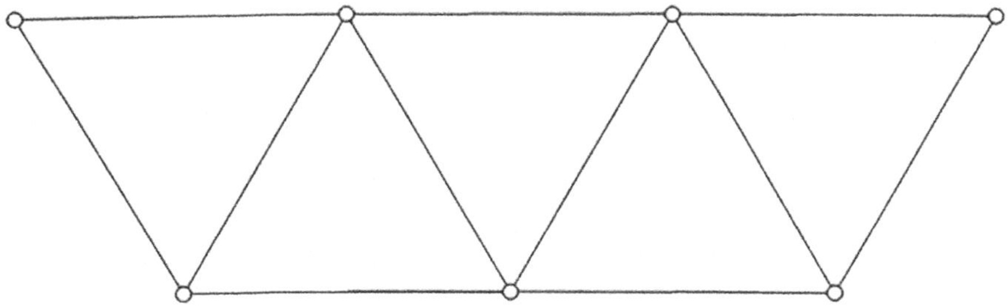

Puzzle 114: Base group $Z_{11}$

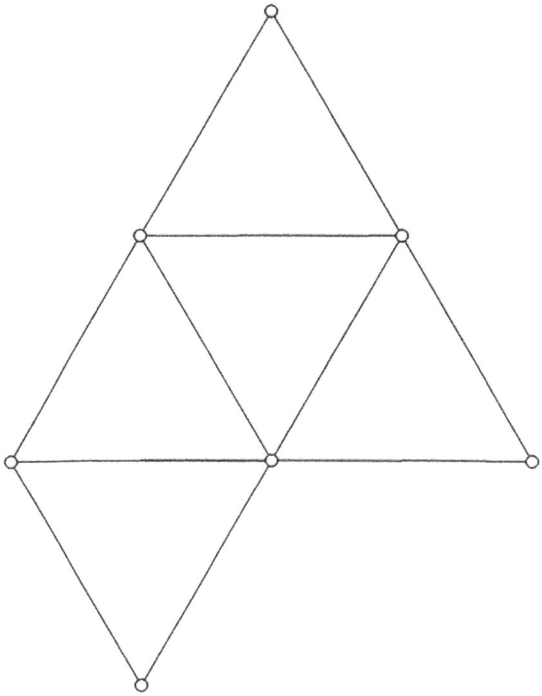

Puzzle 115: Base group $Z_{11}$

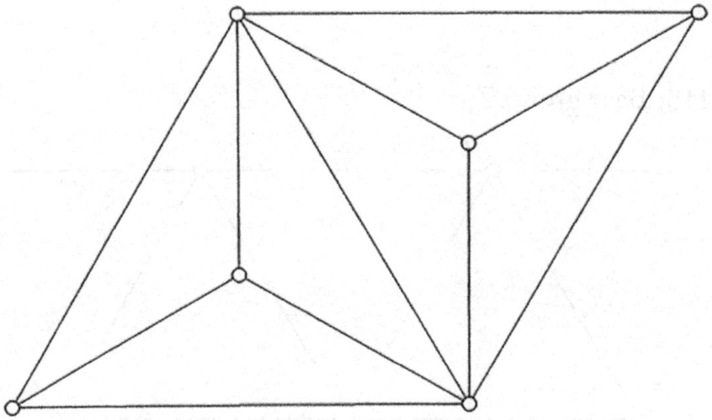

Puzzle 116: Base group $Z_{11}$

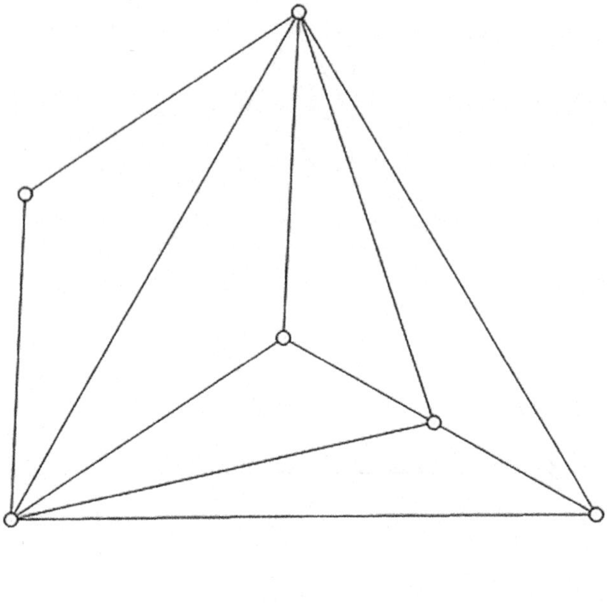

Puzzle 117: Base group $Z_{11}$

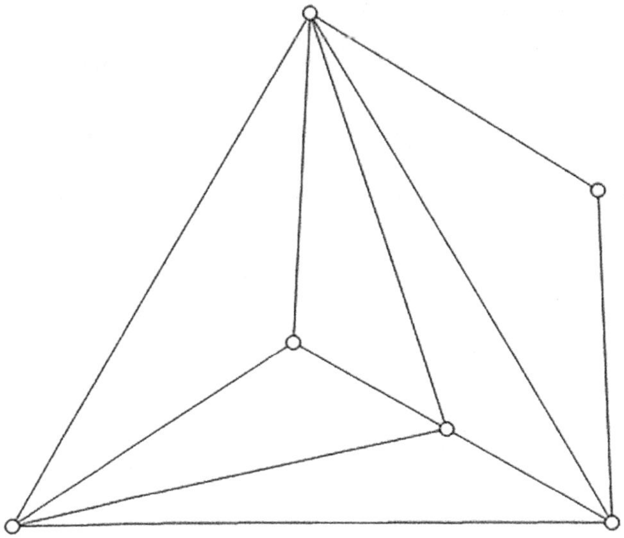

Puzzle 118: Base group $Z_{11}$

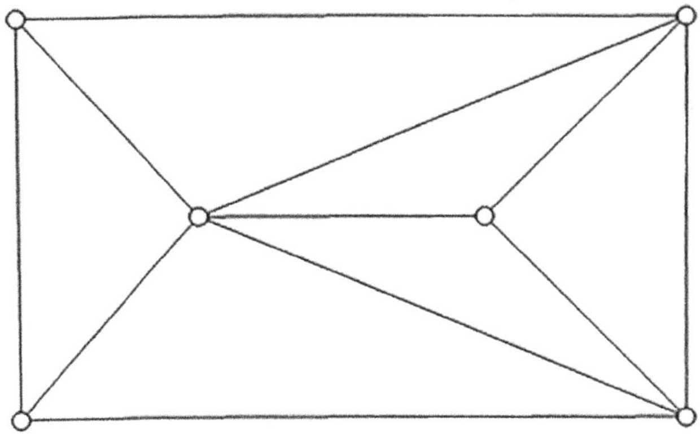

# 15. Triangulations over $Z_{12}$

Puzzle 119: Base group $Z_{12}$

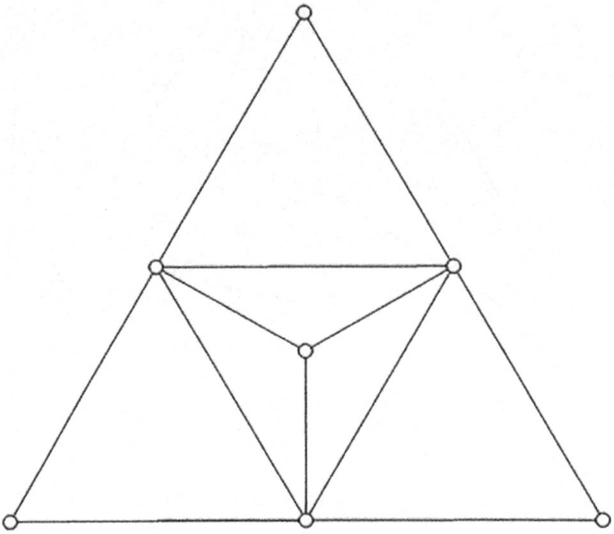

Puzzle 120: Base group $Z_{12}$

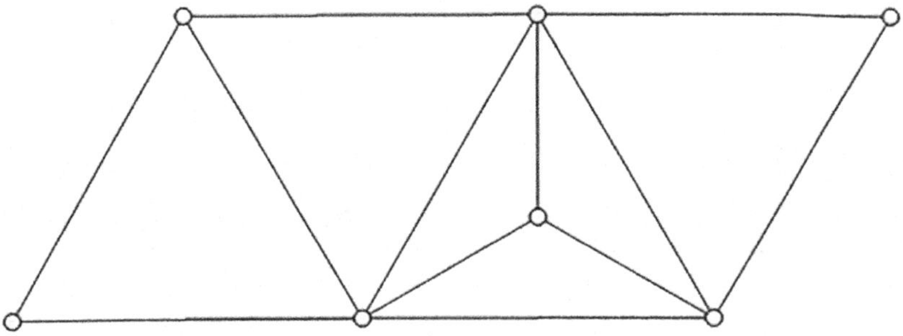

Puzzle 121: Base group $Z_{12}$

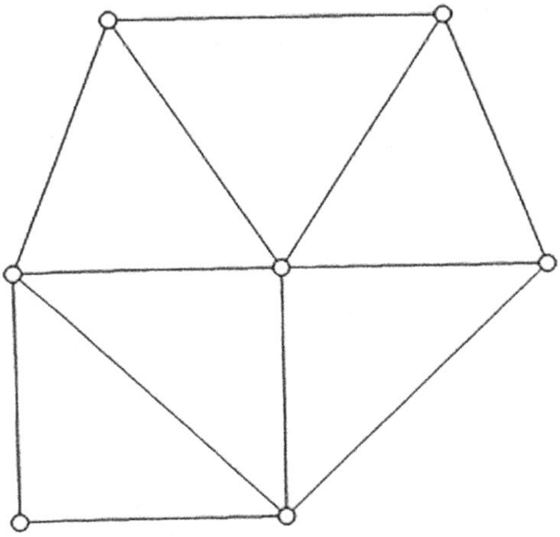

Puzzle 122: Base group $Z_{12}$

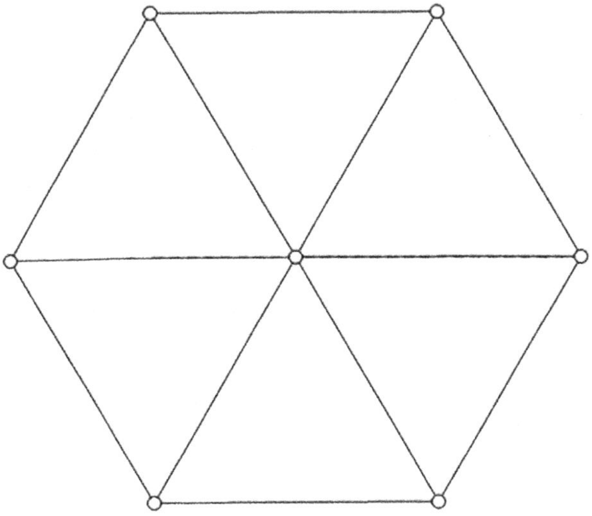

Puzzle 123: Base group $Z_{12}$

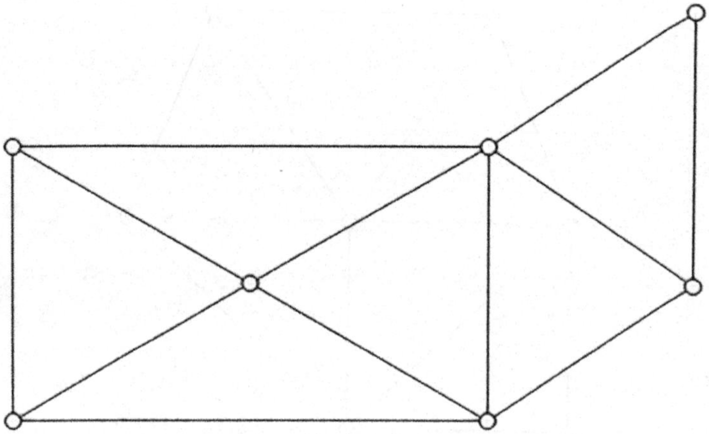

Puzzle 124: Base group $Z_{12}$

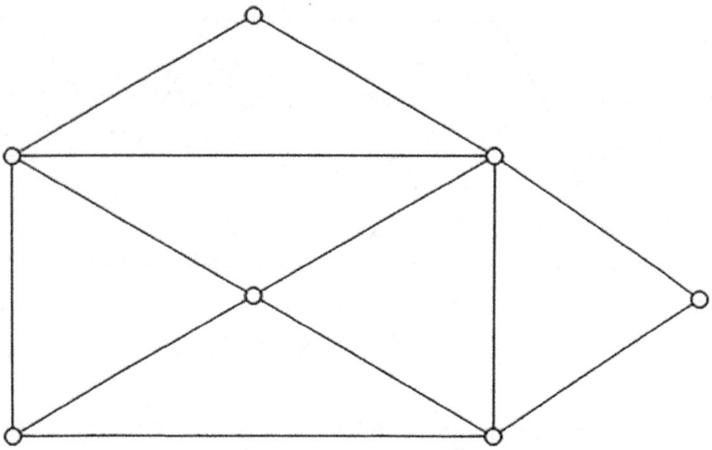

Puzzle 125: Base group $Z_{12}$

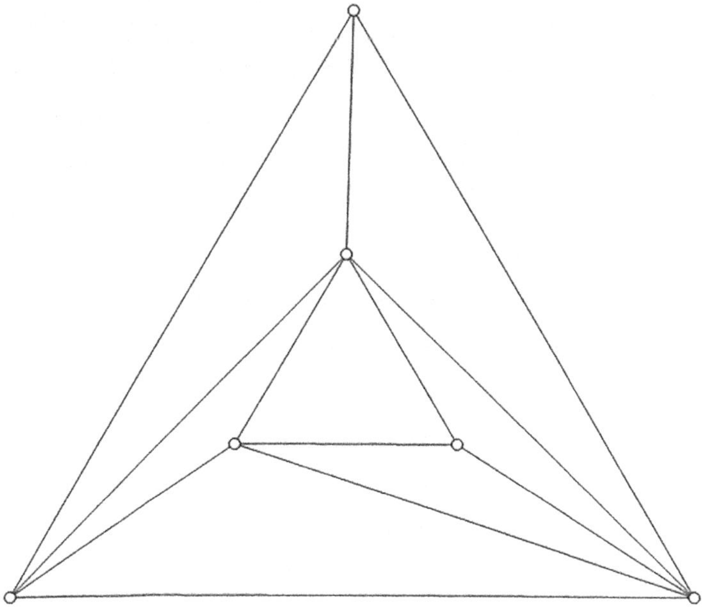

Puzzle 126: Base group $Z_{12}$

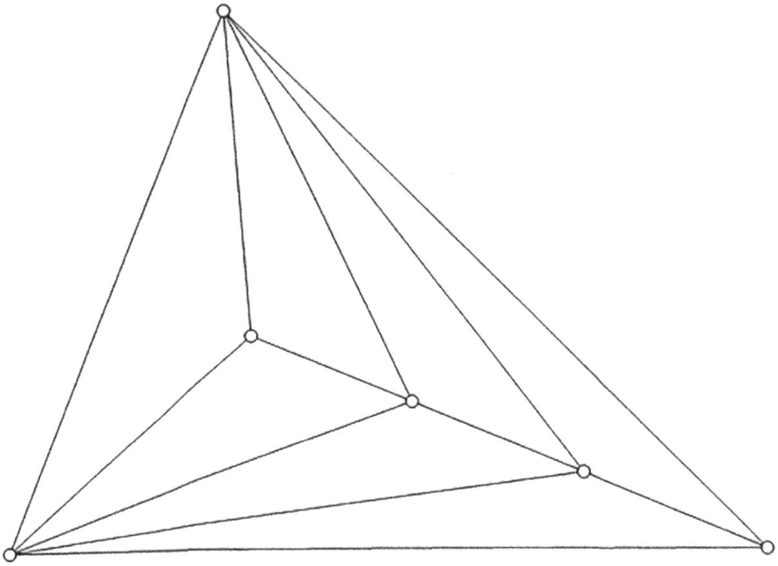

# 16. Triangulations over $Z_{13}$

Puzzle 127: Base group $Z_{13}$

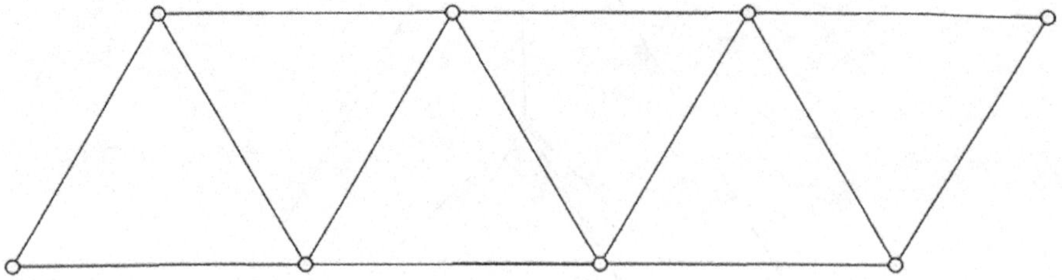

Puzzle 128: Base group $Z_{13}$

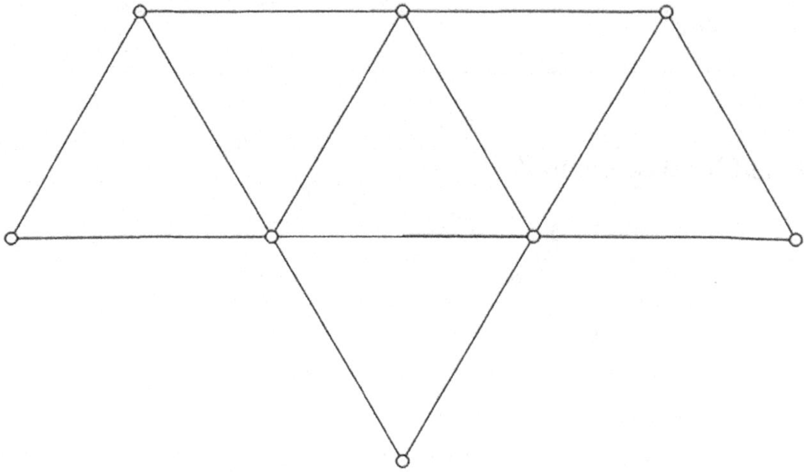

Puzzle 129: Base group $Z_{13}$

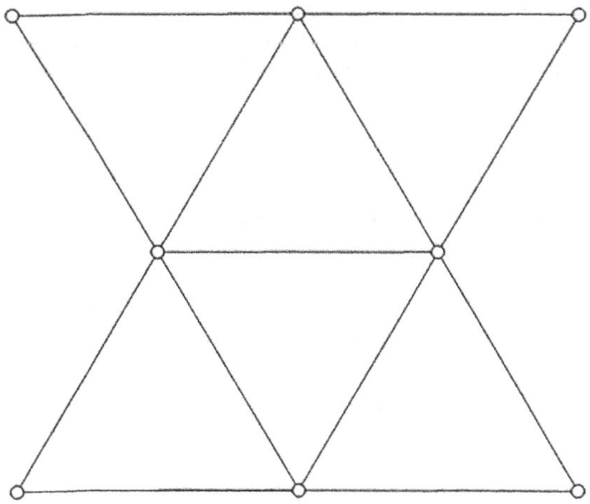

Puzzle 130: Base group $Z_{13}$

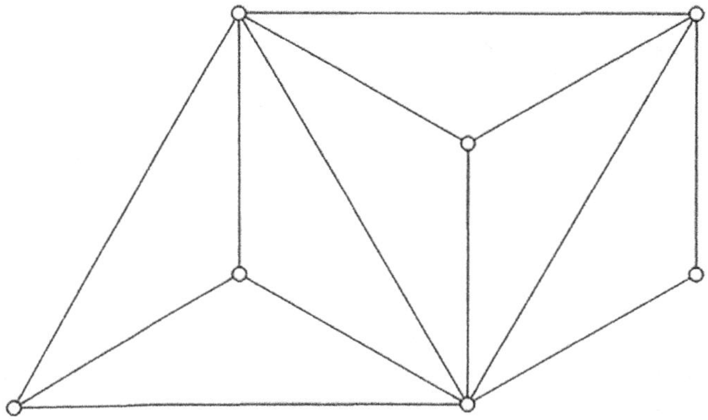

Puzzle 131: Base group $Z_{13}$

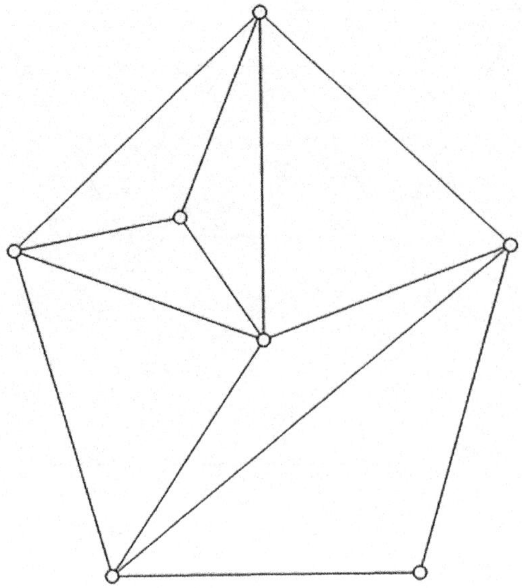

Puzzle 132: Base group $Z_{13}$

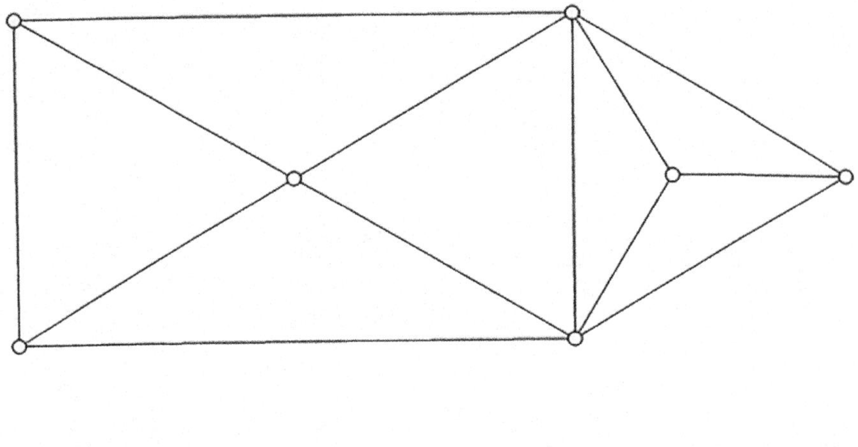

Puzzle 133: Base group $Z_{13}$

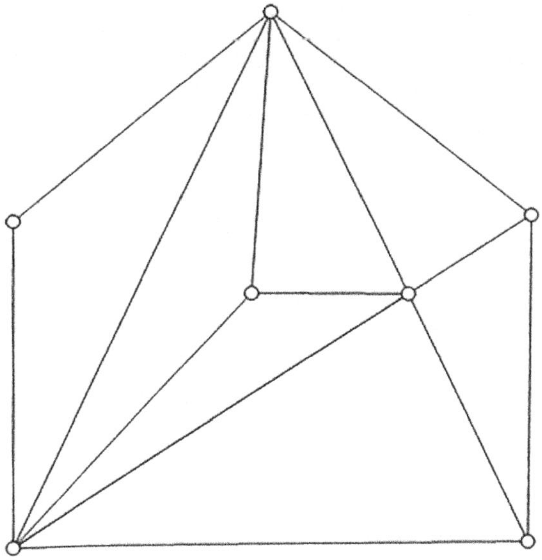

Puzzle 134: Base group $Z_{13}$

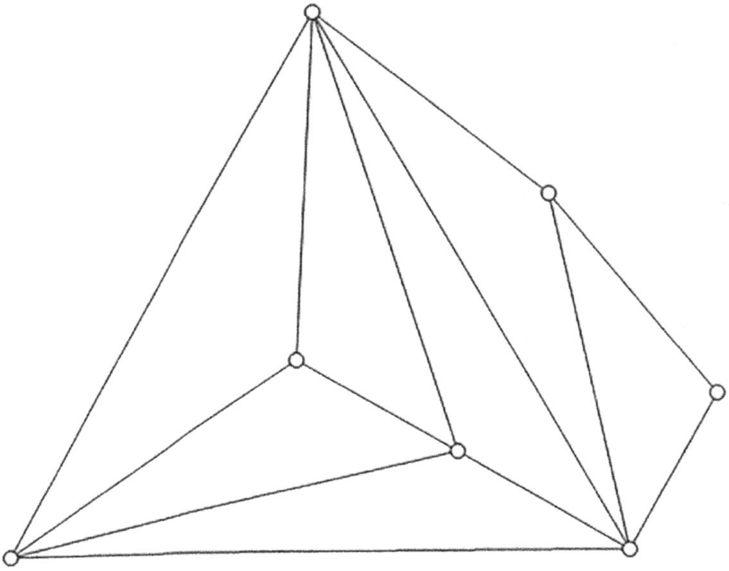

# 17. Triangulations over $Z_{14}$

Puzzle 135: Base group $Z_{14}$

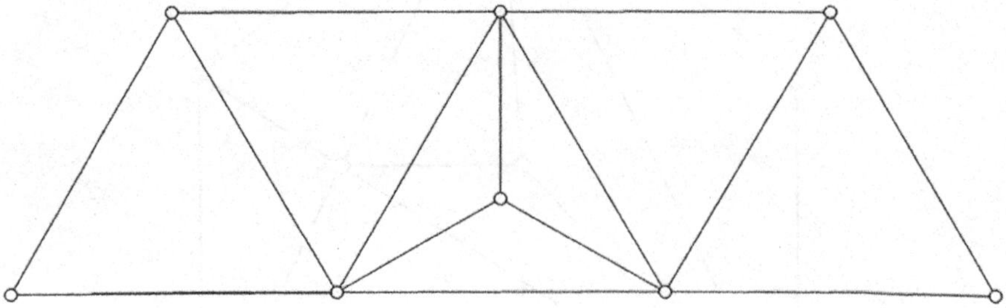

Puzzle 136: Base group $Z_{14}$

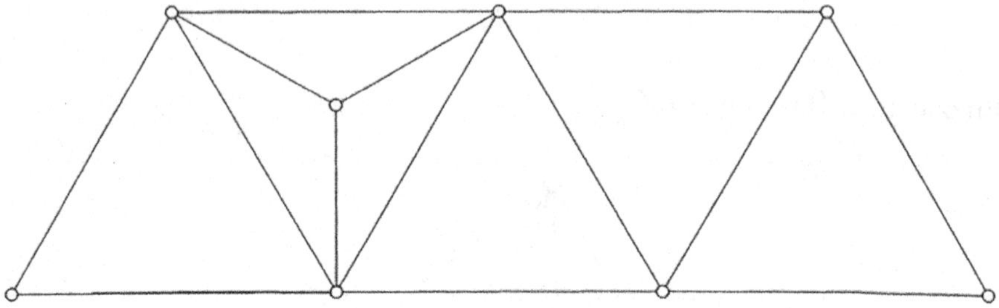

Puzzle 137: Base group $Z_{14}$

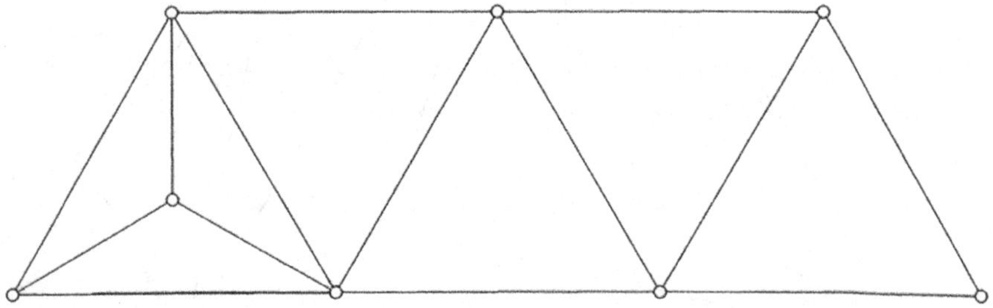

Puzzle 138: Base group $Z_{14}$

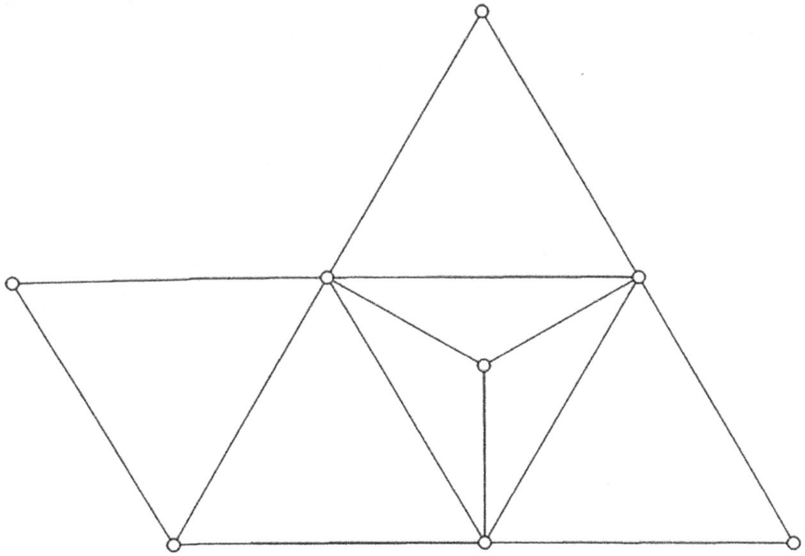

Puzzle 139: Base group $Z_{14}$ ; F = 5

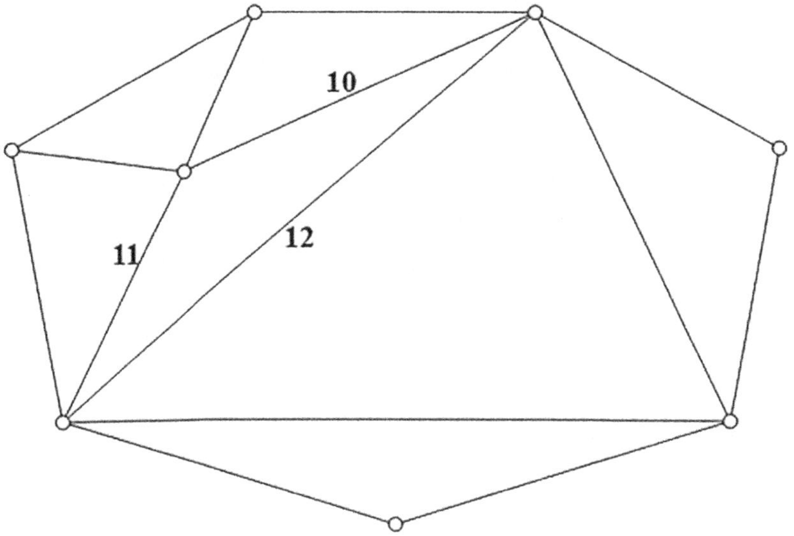

Puzzle 140: Base group $Z_{14}$

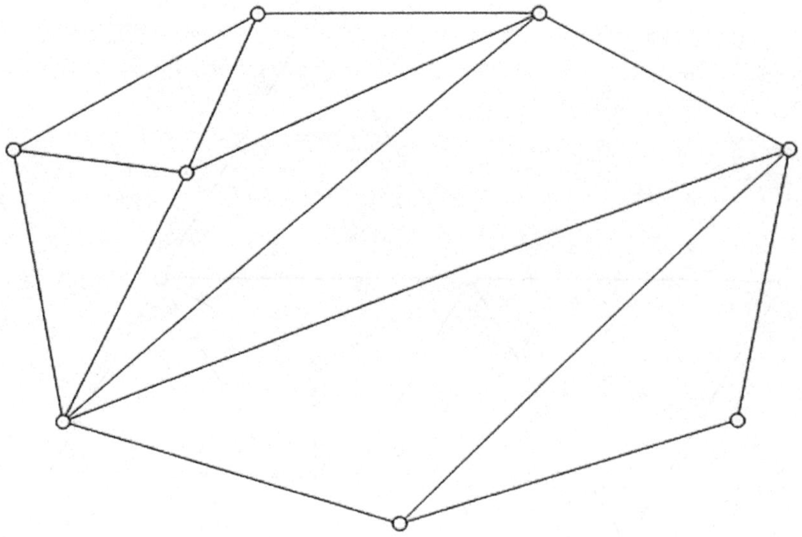

Puzzle 141: Base group $Z_{14}$

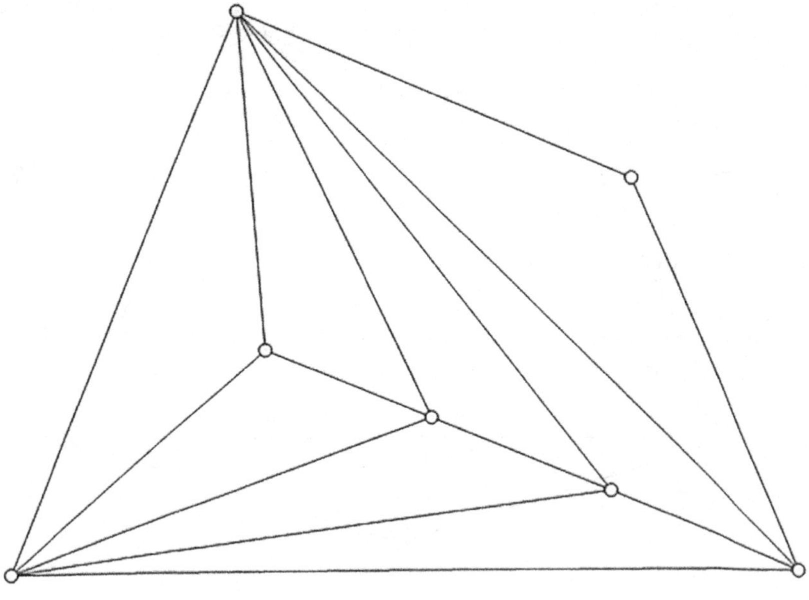

Puzzle 142: Base group $Z_{14}$

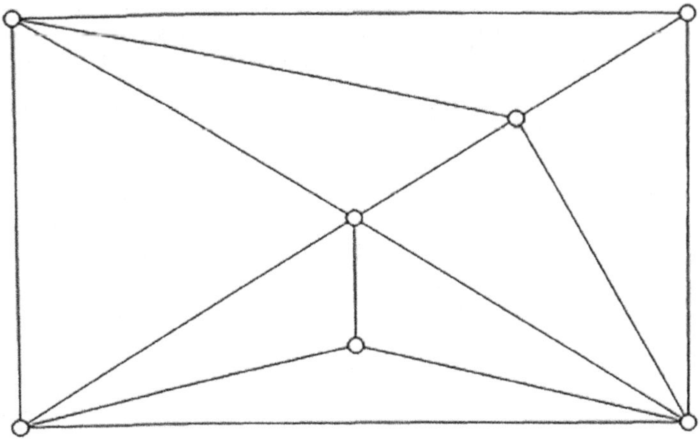

Puzzle 143: Base group $Z_{14}$ ; F = 5

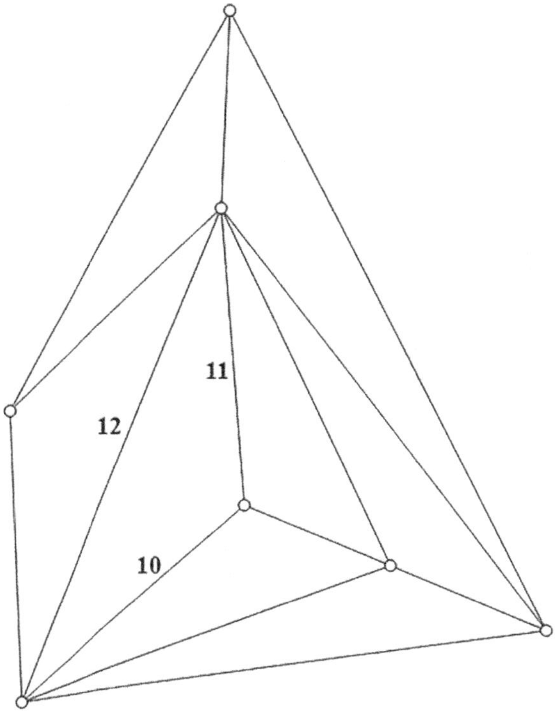

Puzzle 144: Base group $Z_{14}$

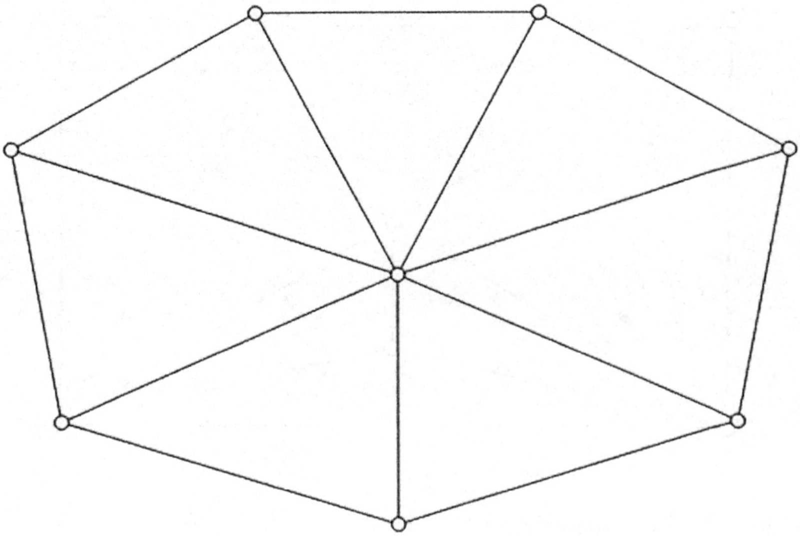

# 18. Triangulations over $Z_{15}$

Puzzle 145: Base group $Z_{15}$

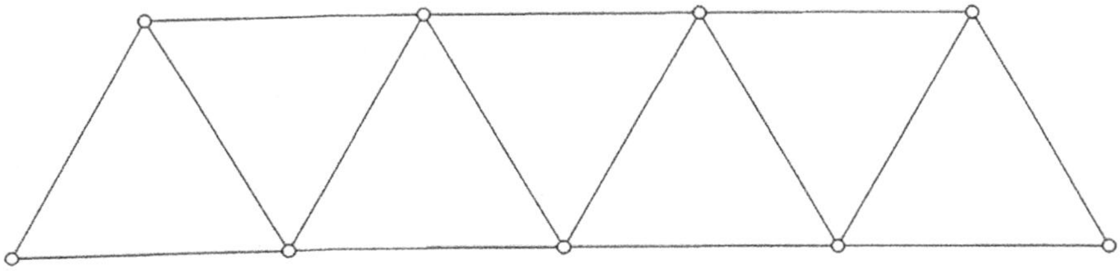

Puzzle 146: Base group $Z_{15}$

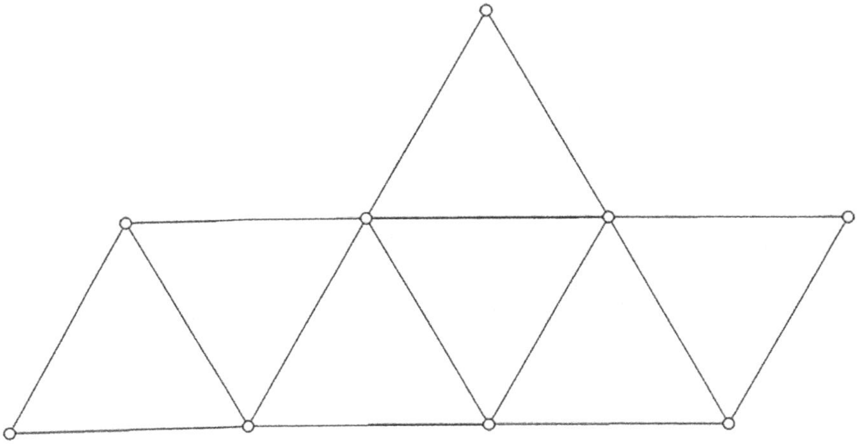

Puzzle 147: Base group $Z_{15}$

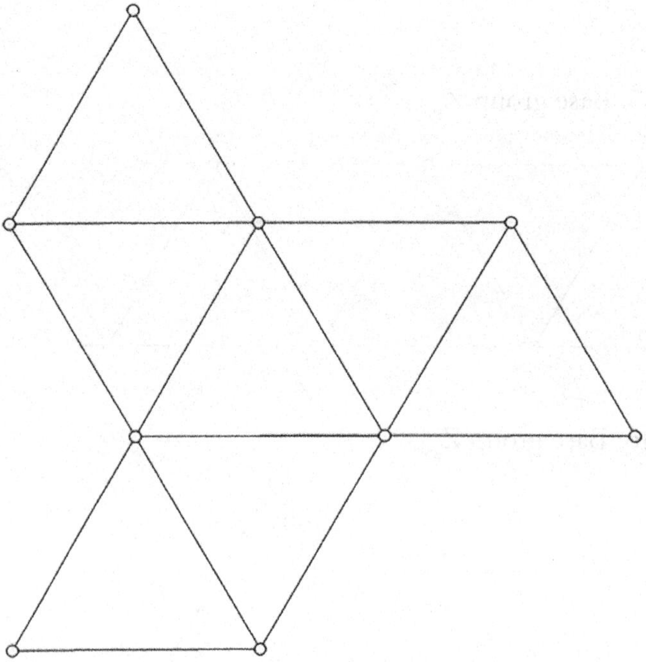

Puzzle 148: Base group $Z_{15}$

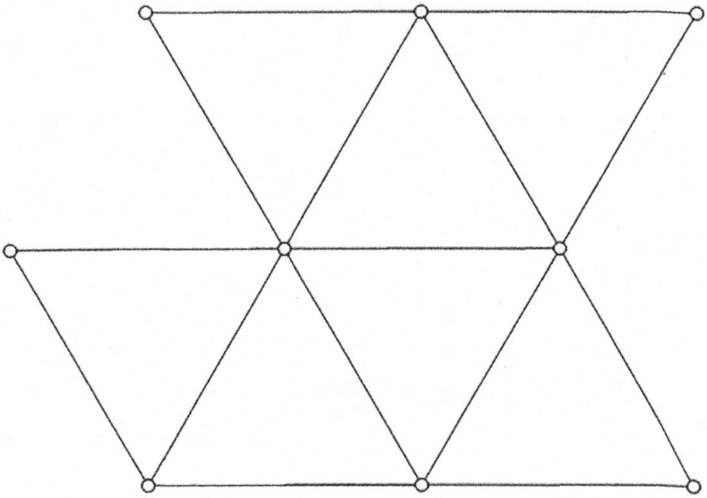

Puzzle 149: Base group $Z_{15}$

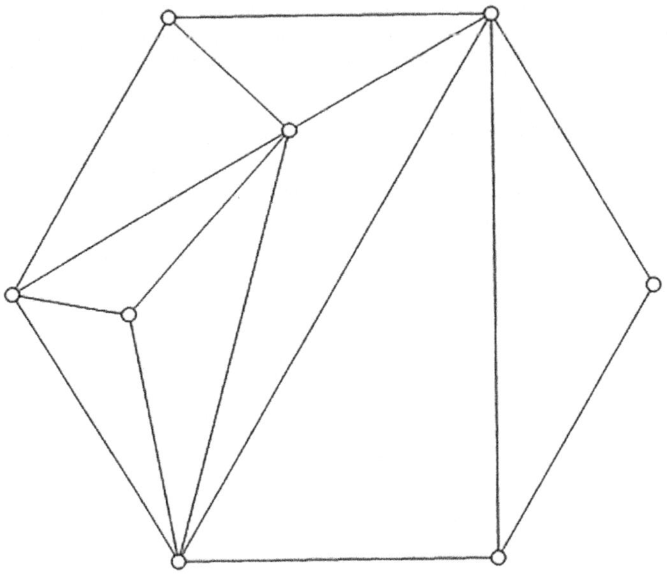

Puzzle 150: Base group $Z_{15}$

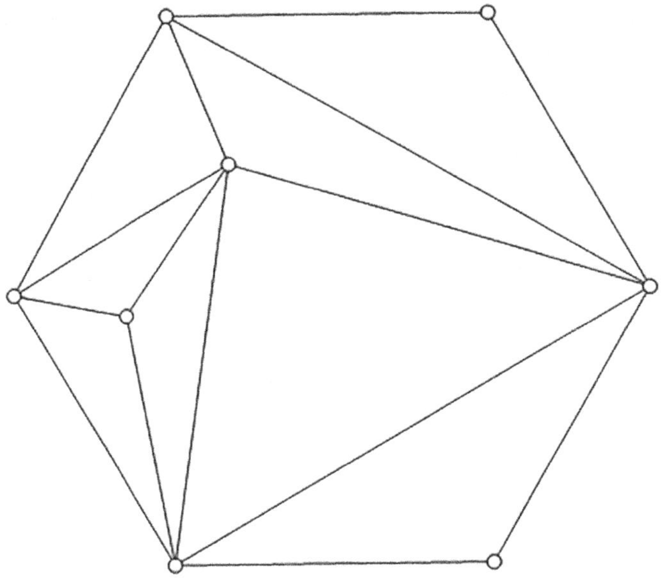

Puzzle 151: Base group $Z_{15}$

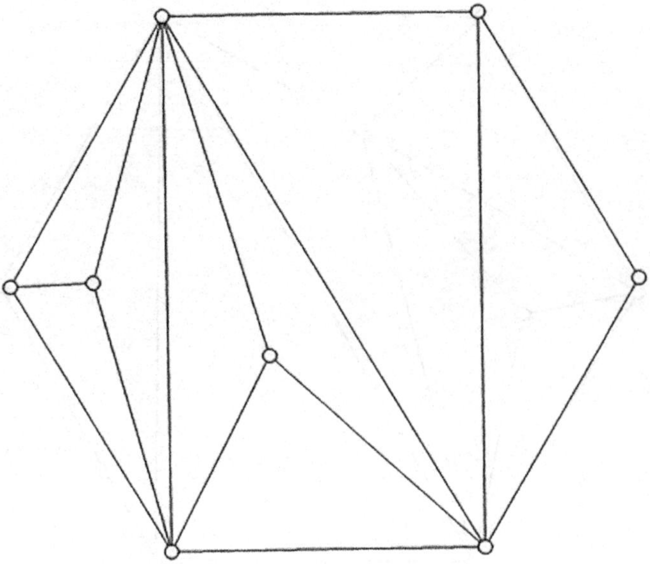

Puzzle 152: Base group $Z_{15}$

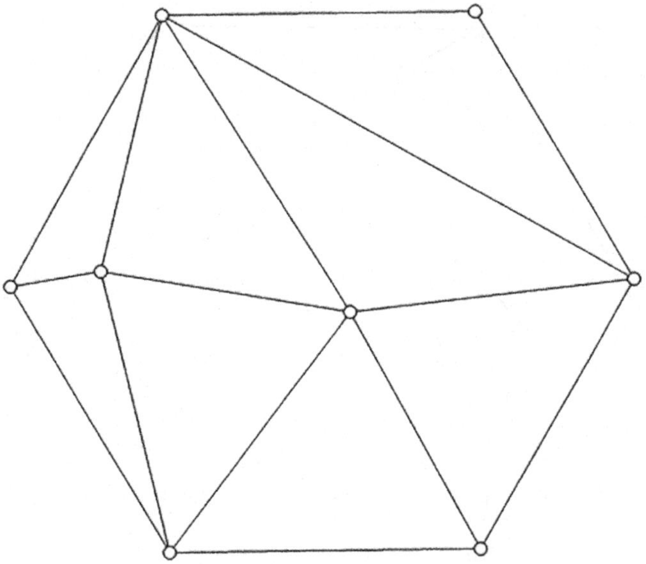

Puzzle 153: Base group $Z_{15}$

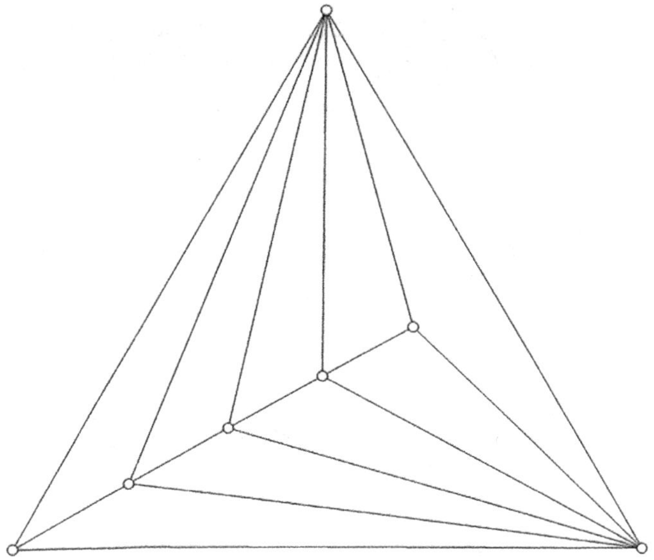

Puzzle 154: Base group $Z_{15}$

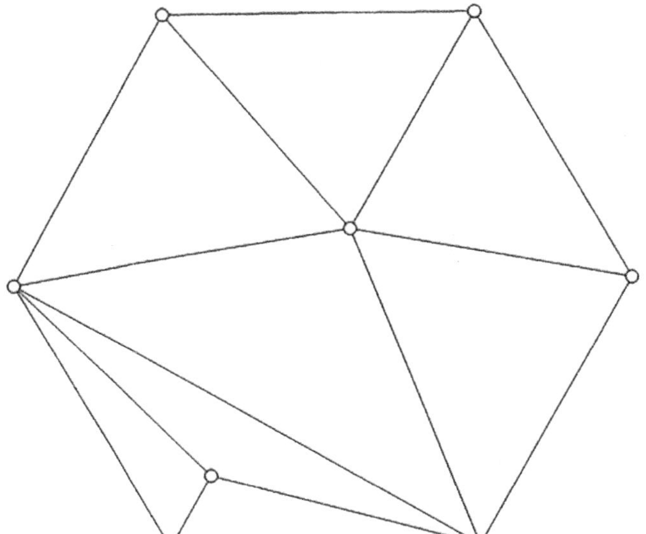

Puzzle 155: Base group $Z_{15}$

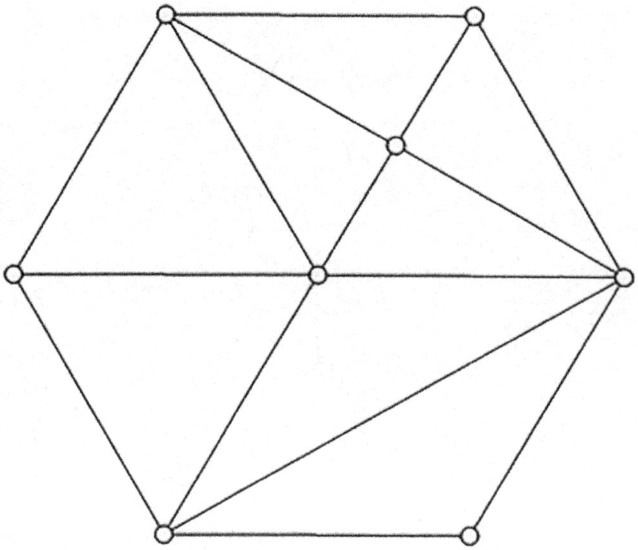

Puzzle 156: Base group $Z_{15}$

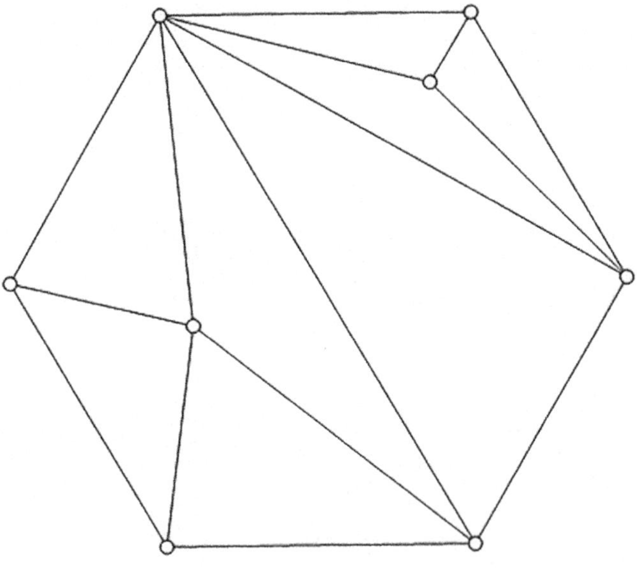

Puzzle 157: Base group $Z_{15}$

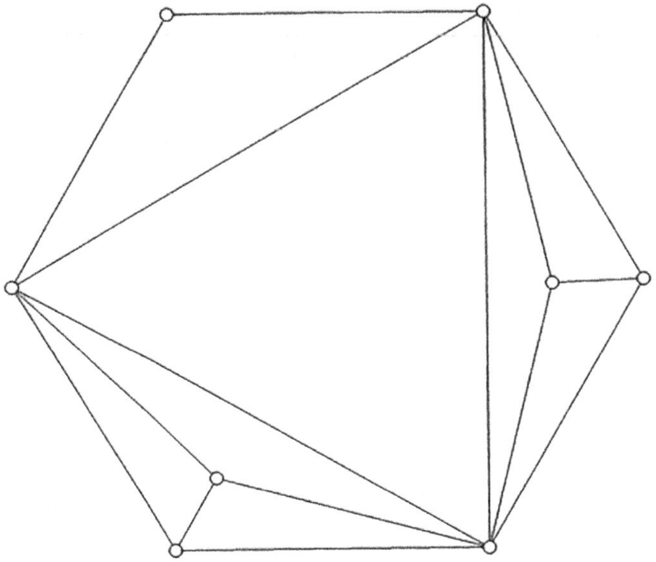

Puzzle 158: Base group $Z_{15}$

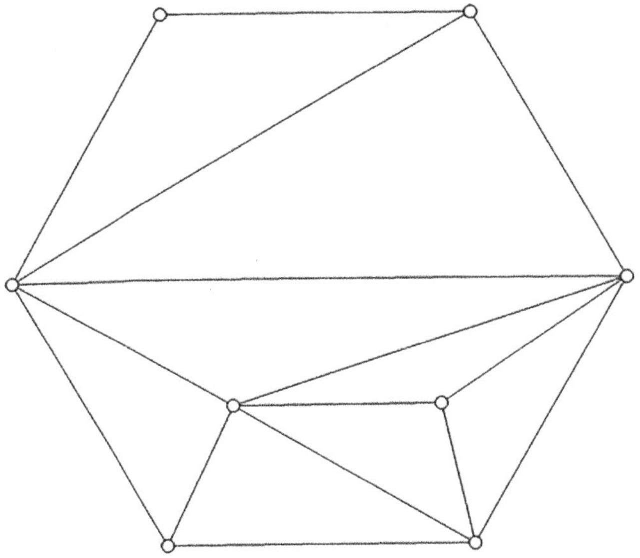

Puzzle 159: Base group $Z_{15}$

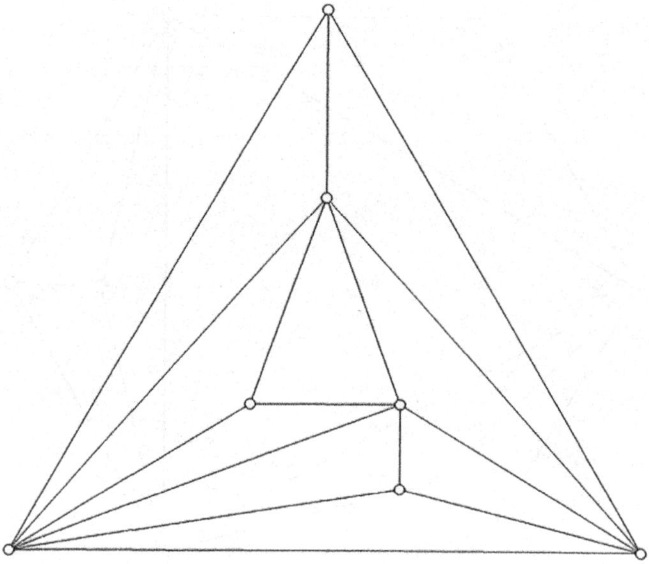

Puzzle 160: Base group $Z_{15}$

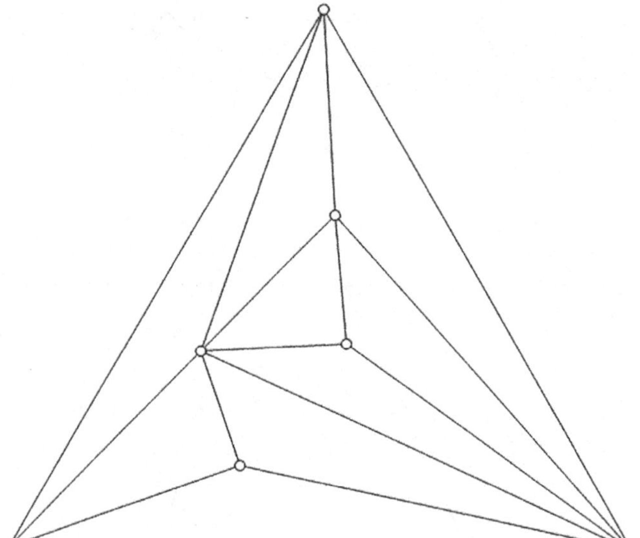

# 19. Triangulations over $Z_{16}$

Puzzle 161: Base group $Z_{16}$

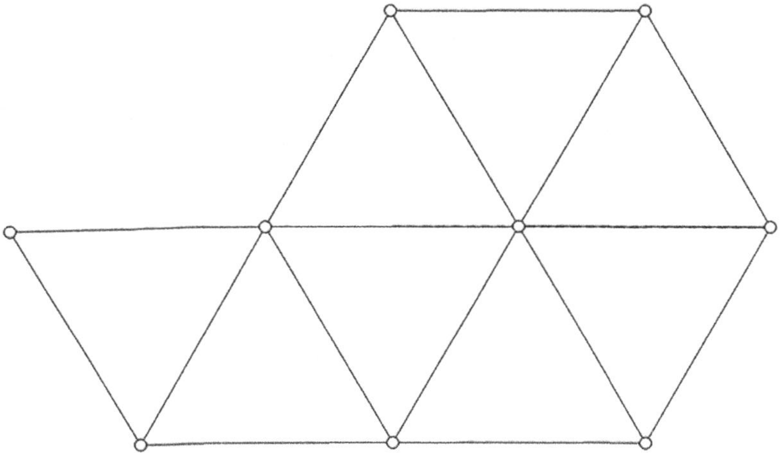

Puzzle 162: Base group $Z_{16}$

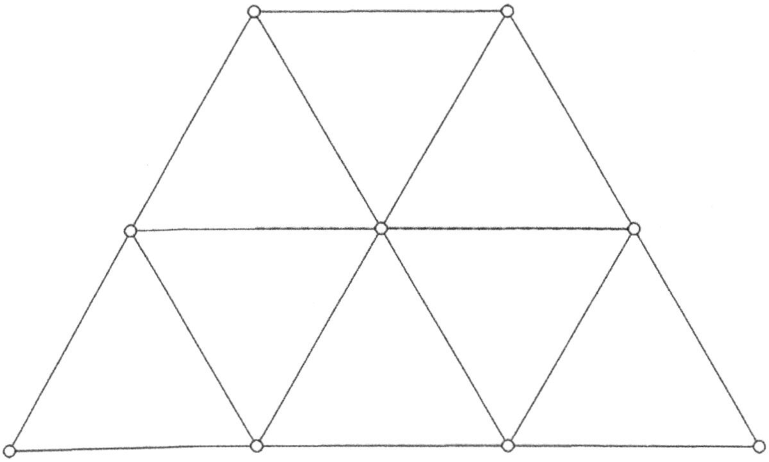

Puzzle 163: Base group $Z_{16}$

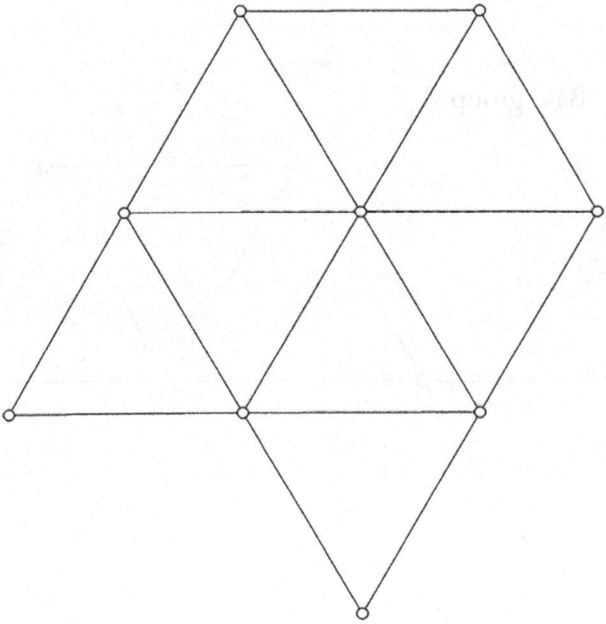

Puzzle 164: Base group $Z_{16}$

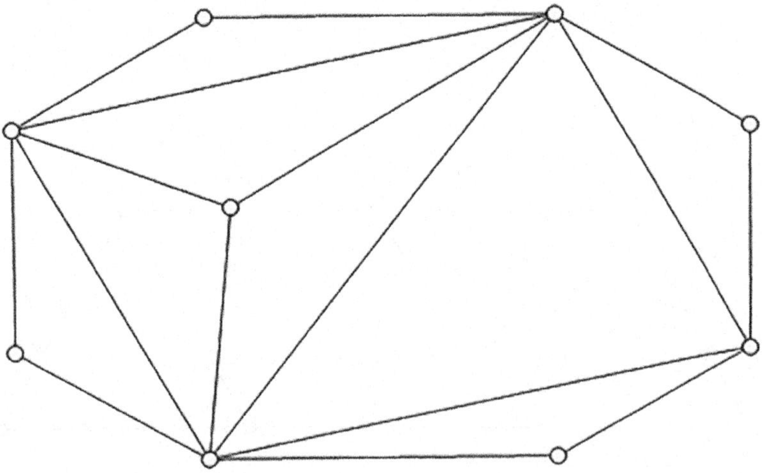

Puzzle 165: Base group $Z_{16}$

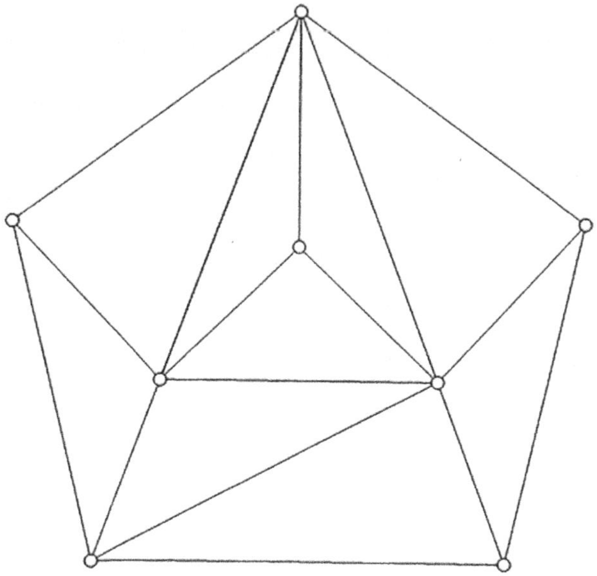

Puzzle 166: Base group $Z_{16}$

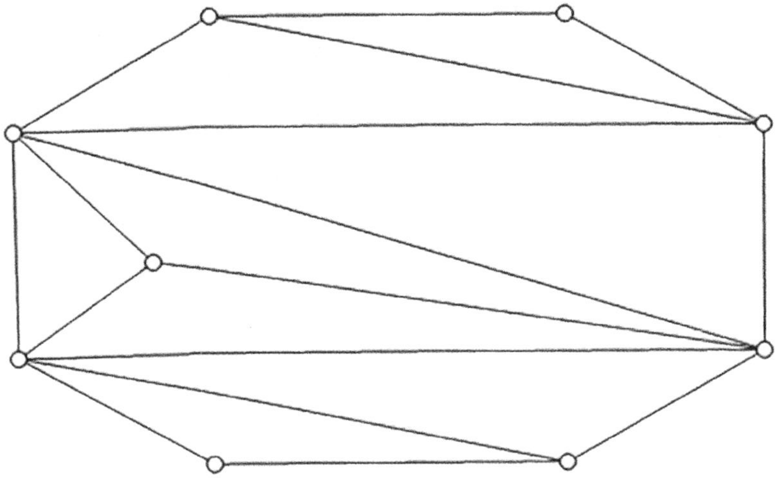

Puzzle 167: Base group $Z_{16}$

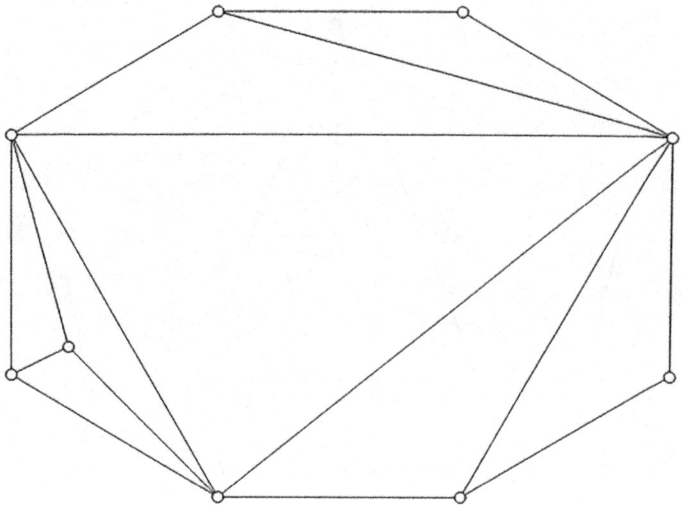

Puzzle 168: Base group $Z_{16}$

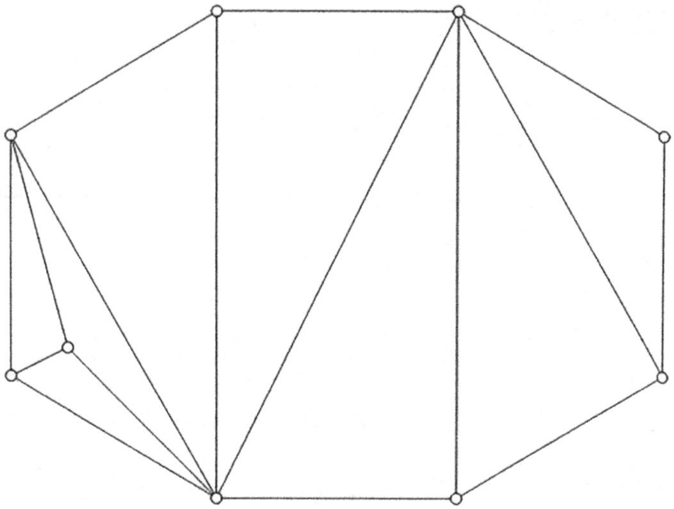

Puzzle 169: Base group $Z_{16}$

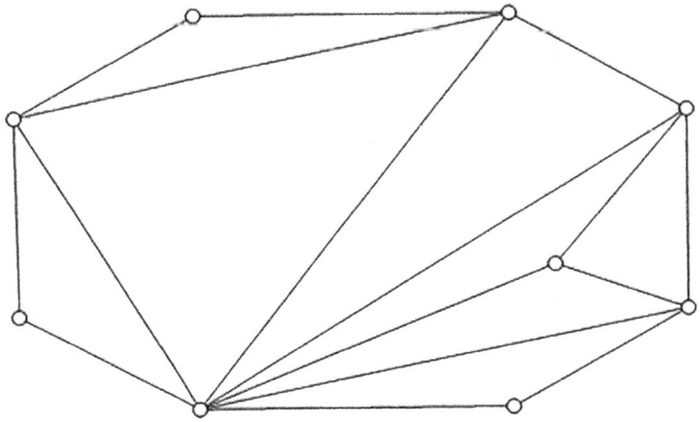

Puzzle 170: Base group $Z_{16}$

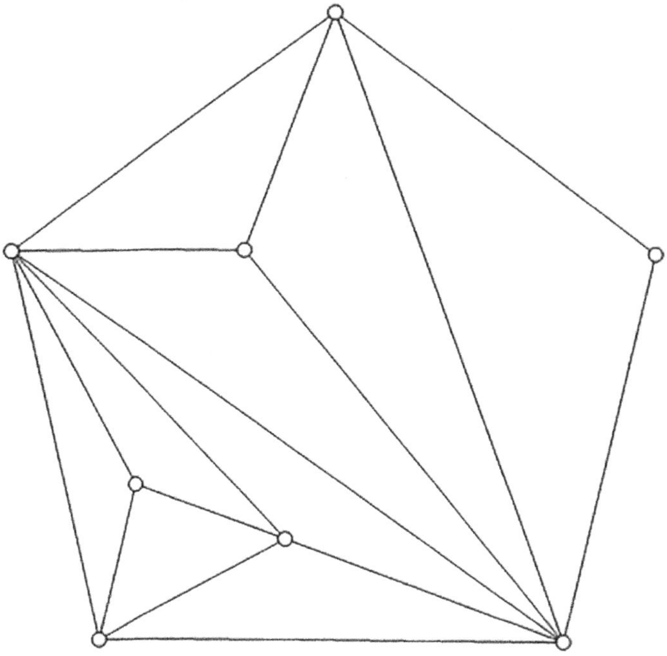

Puzzle 171: Base group $Z_{16}$

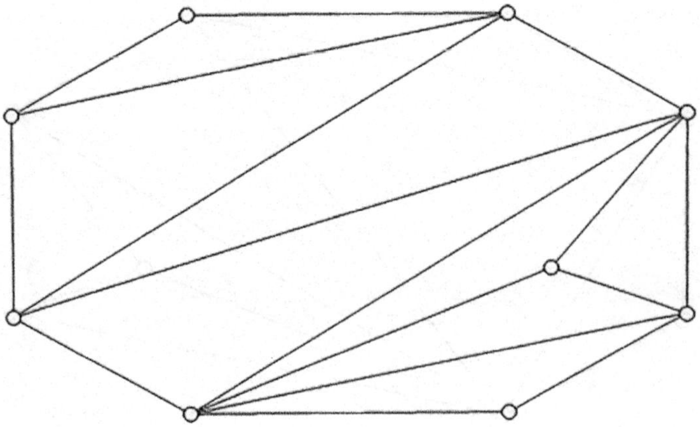

Puzzle 172: Base group $Z_{16}$

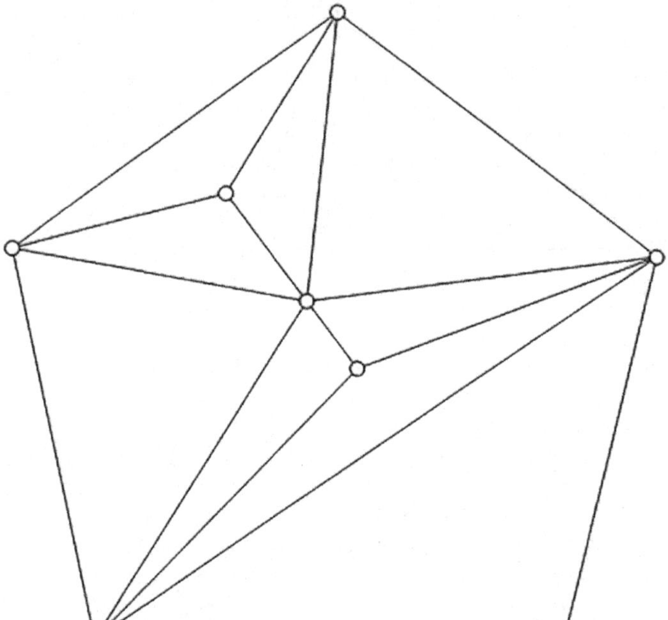

Puzzle 173: Base group $Z_{16}$

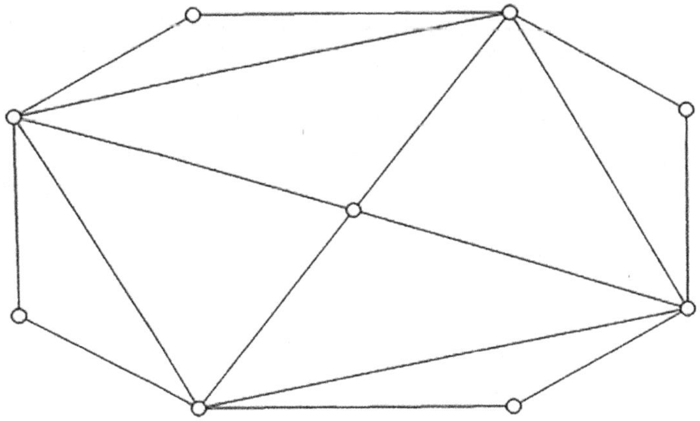

Puzzle 174: Base group $Z_{16}$

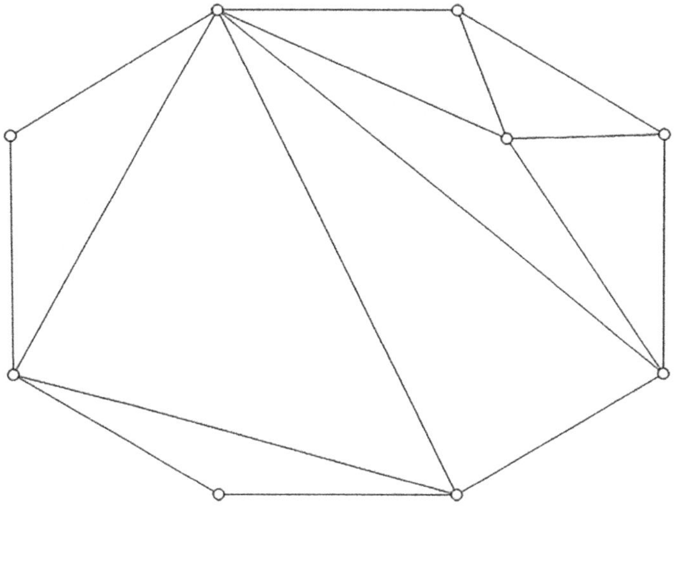

Puzzle 175: Base group $Z_{16}$

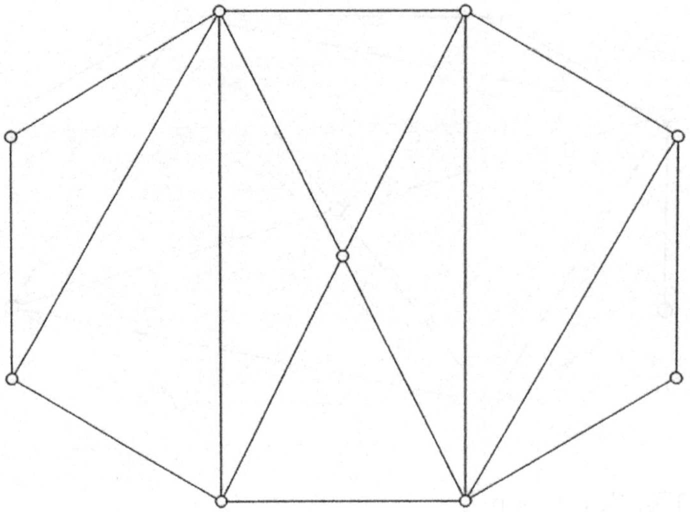

Puzzle 176: Base group $Z_{16}$

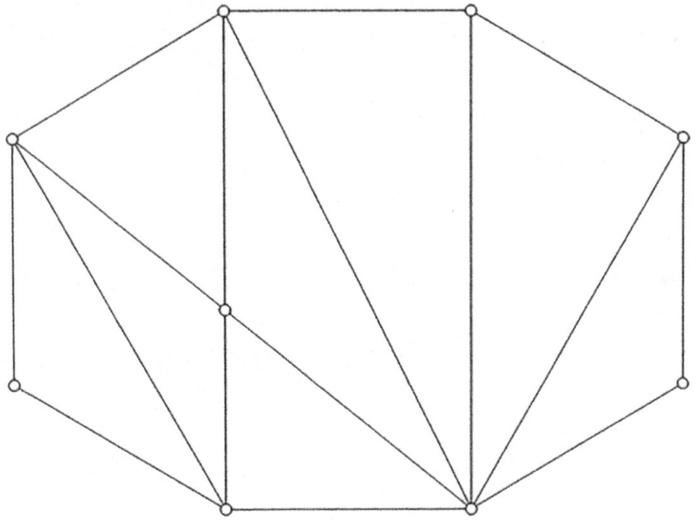

Puzzle 177: Base group $Z_{16}$

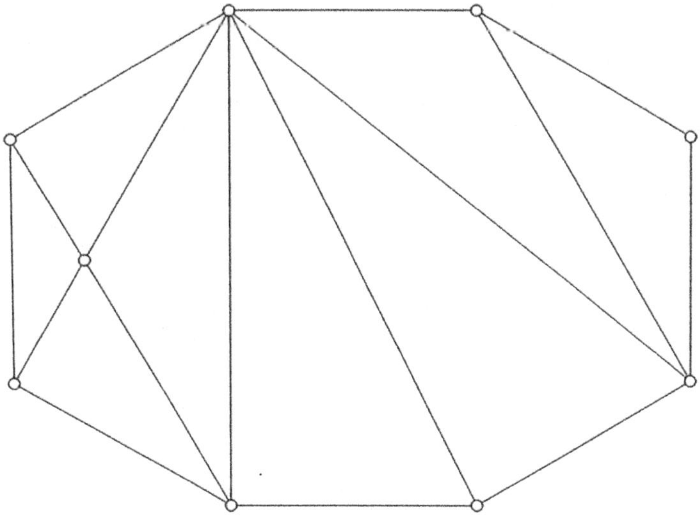

Puzzle 178: Base group $Z_{16}$

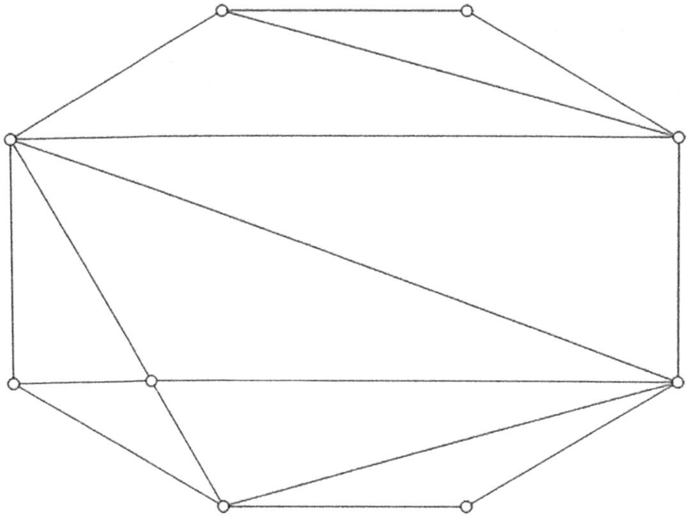

Puzzle 179: Base group $Z_{16}$

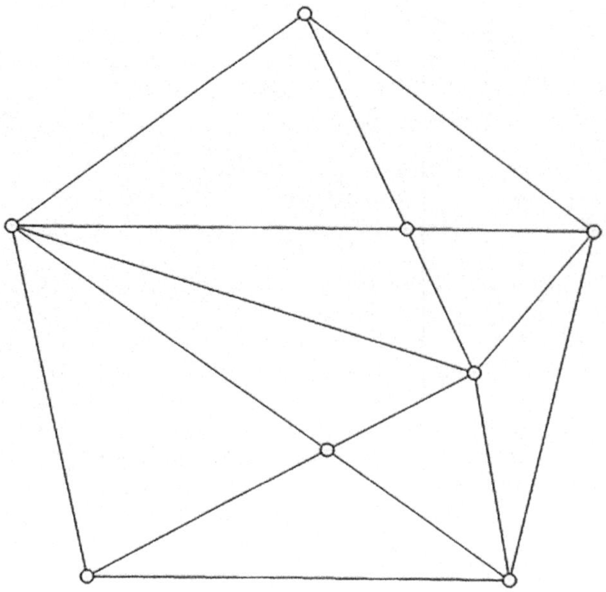

Puzzle 180: Base group $Z_{16}$

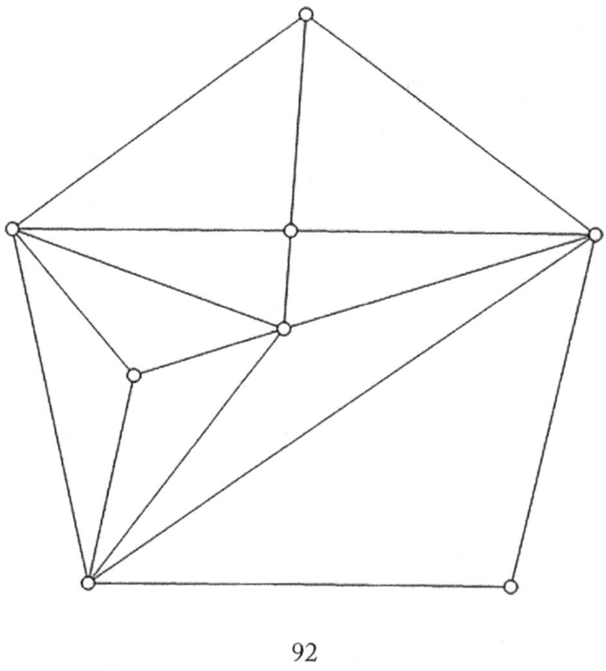

Puzzle 181: Base group $Z_{16}$

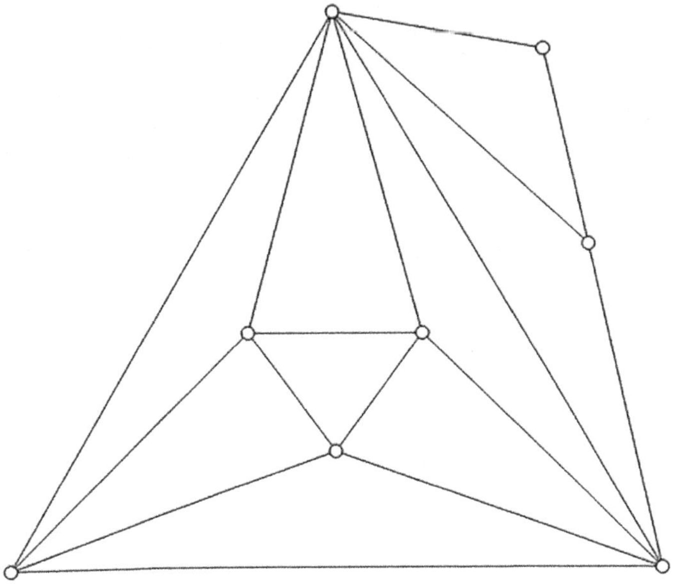

Puzzle 182: Base group $Z_{16}$

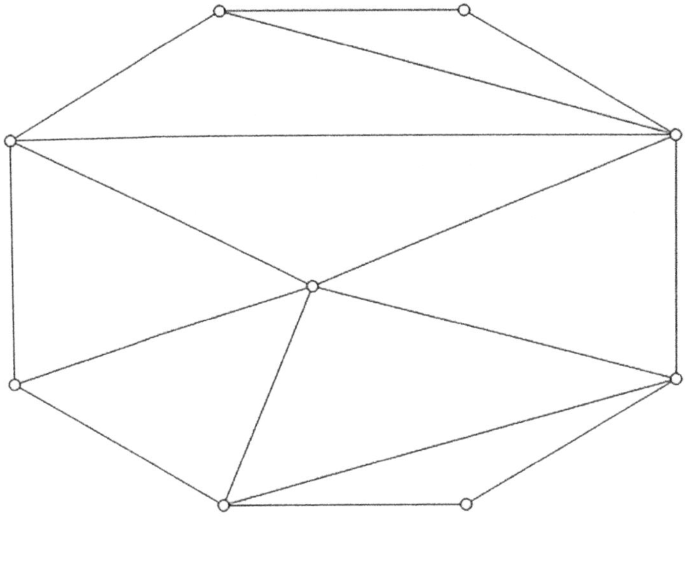

Puzzle 183: Base group $Z_{16}$

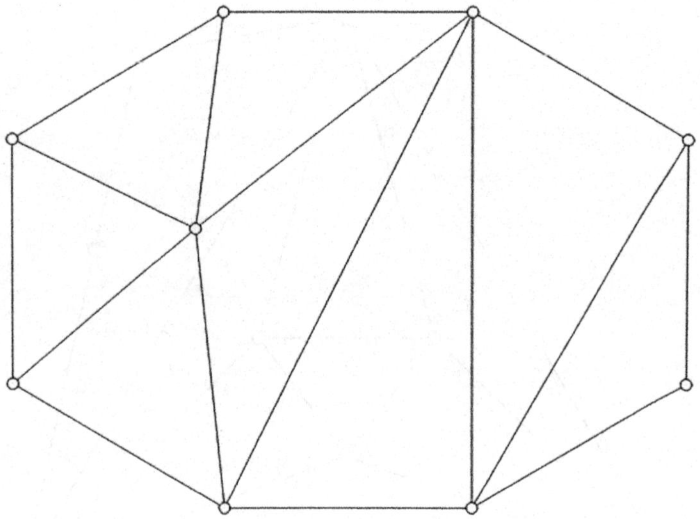

Puzzle 184: Base group $Z_{16}$

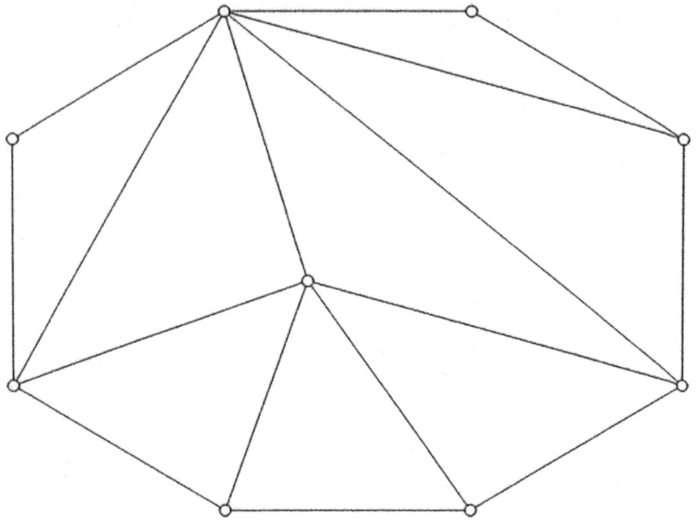

Puzzle 185: Base group $Z_{16}$

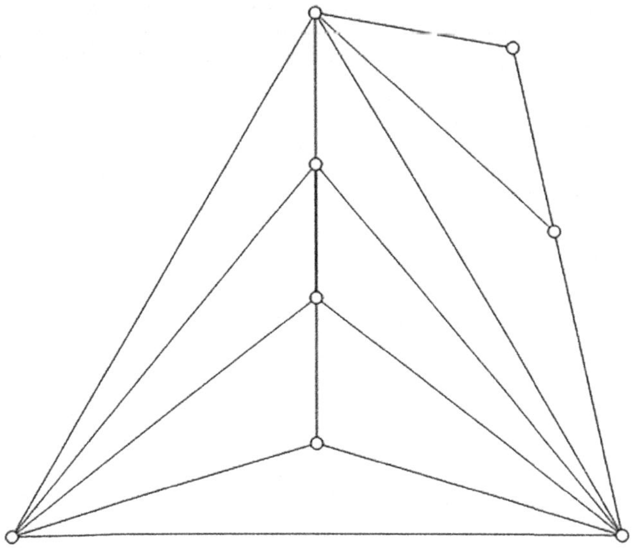

Puzzle 186: Base group $Z_{16}$

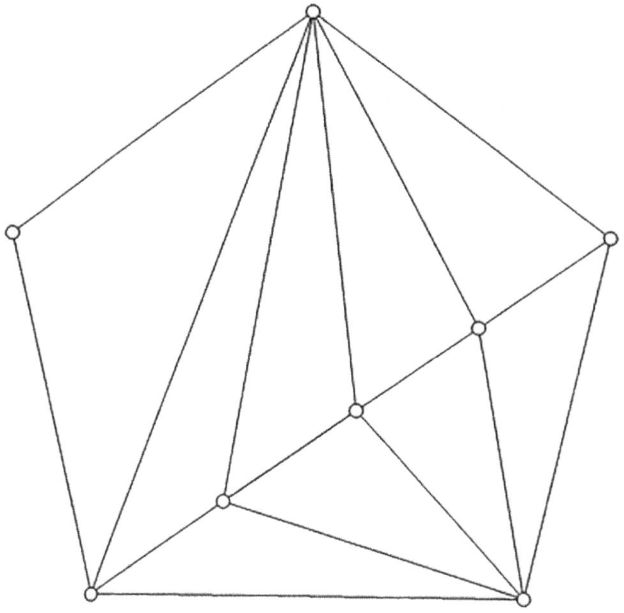

# 20. Linear triangulations

(See also puzzles 102, 105, 113, 127, 145)

Puzzle 187: Base group $Z_{17}$

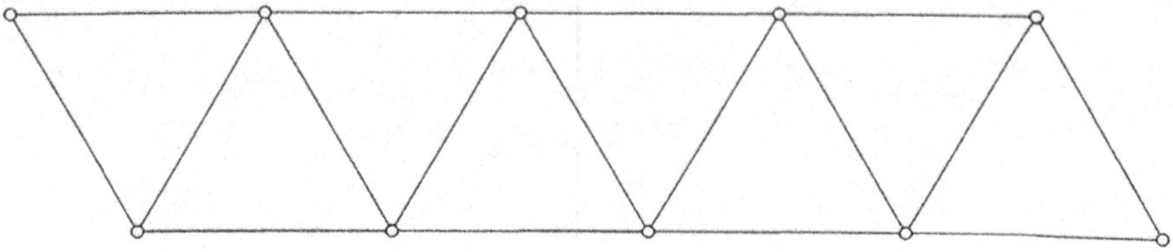

Puzzle 188: Base group $Z_{19}$

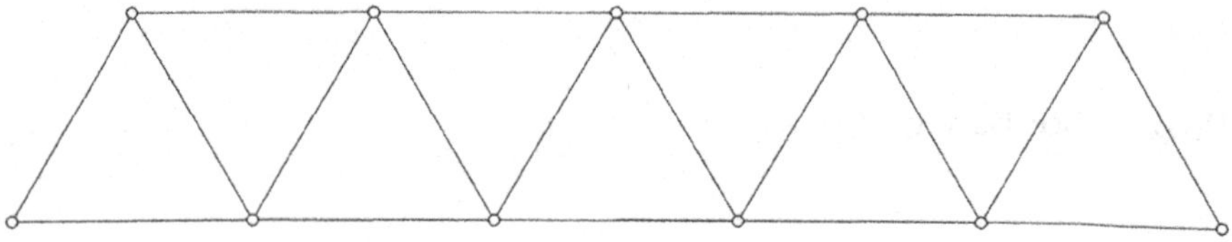

Puzzle 189: Base group $Z_{21}$

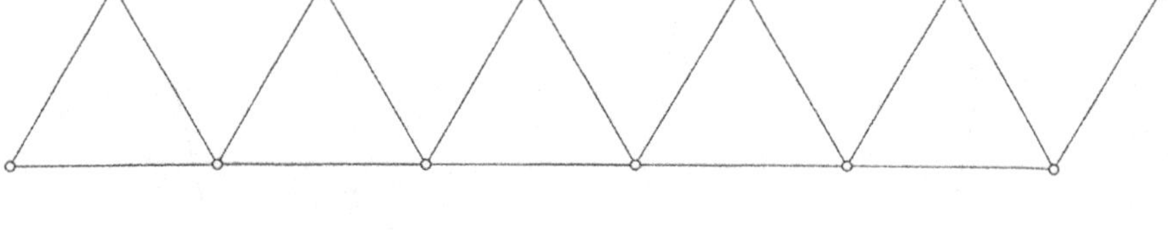

# 21. Equilateral tree triangulations

(These include the linear triangulations in Section 20 as well as puzzles 106, 114, 129, 146, 147, 148, 149.)

Puzzle 190: Base group $Z_{17}$

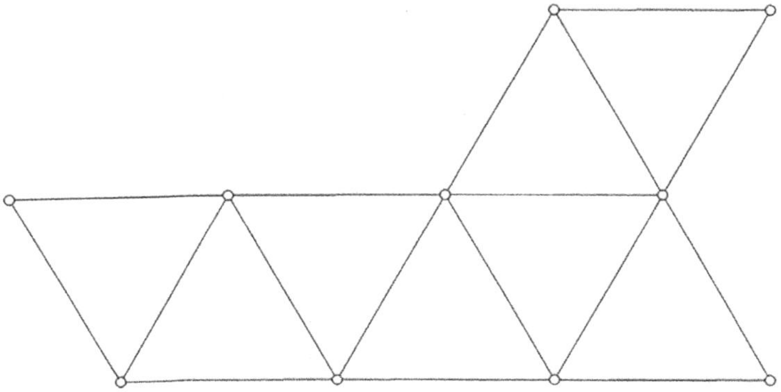

Puzzle 191: Base group $Z_{17}$

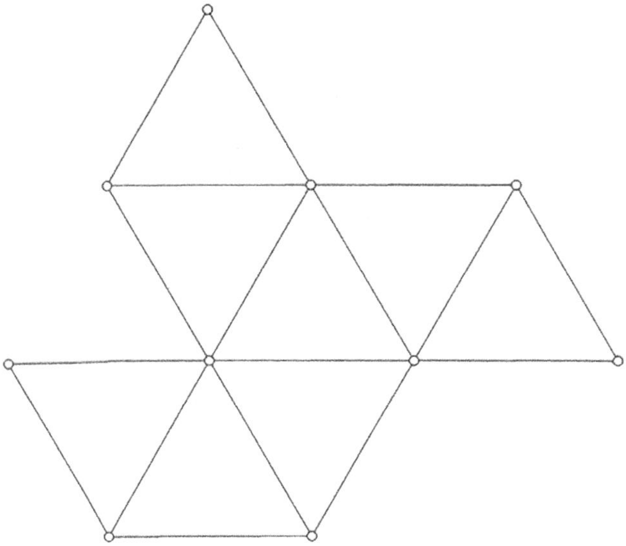

Puzzle 192: Base group $Z_{17}$

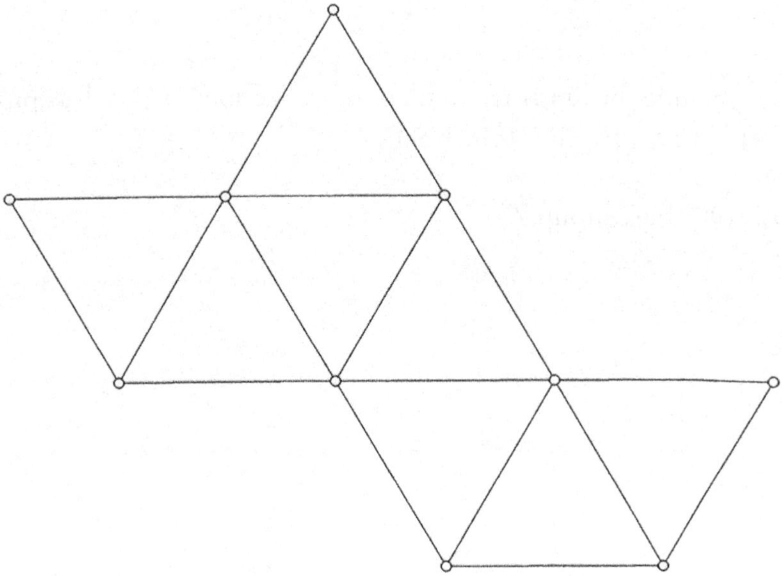

Puzzle 193: Base group $Z_{19}$

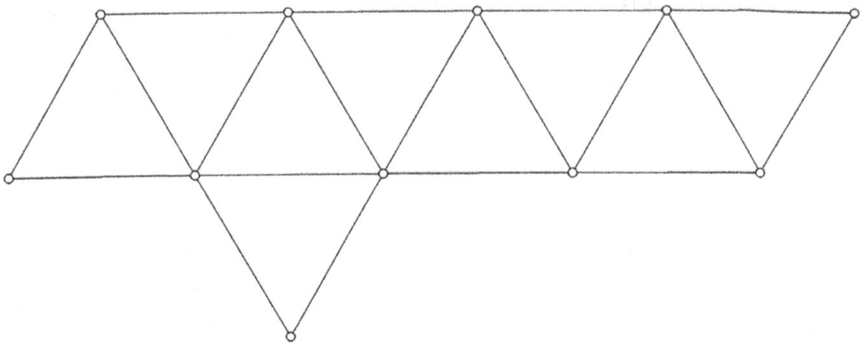

Puzzle 194: Base group $Z_{19}$

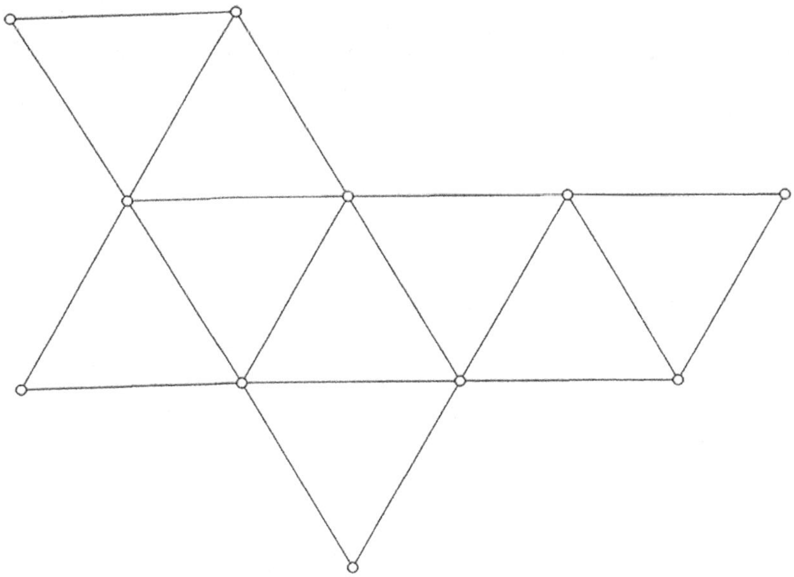

Puzzle 195: Base group $Z_{19}$

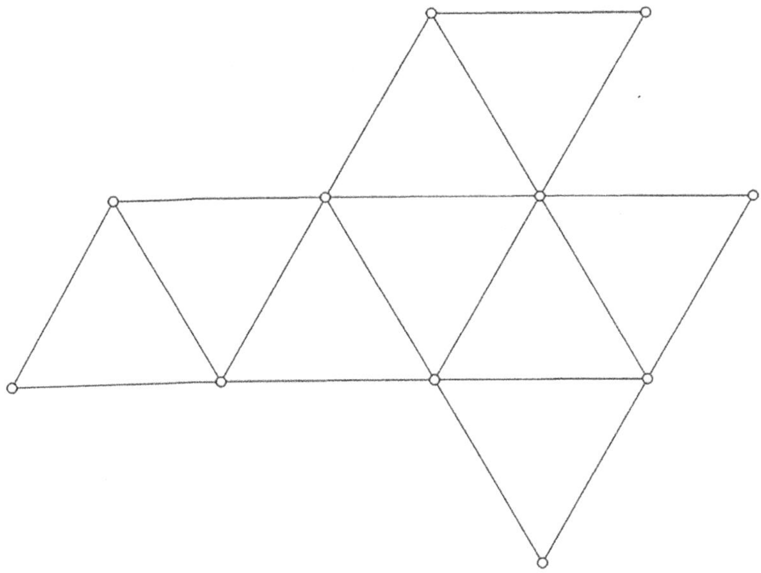

Puzzle 196: Base group $Z_{19}$

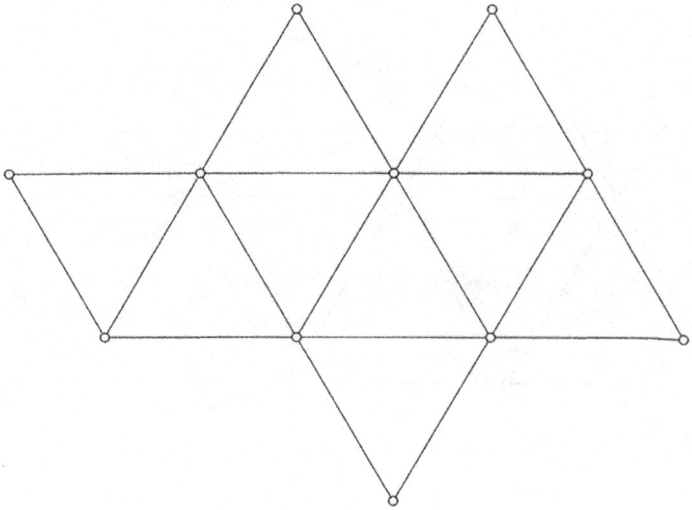

Puzzle 197: Base group $Z_{21}$

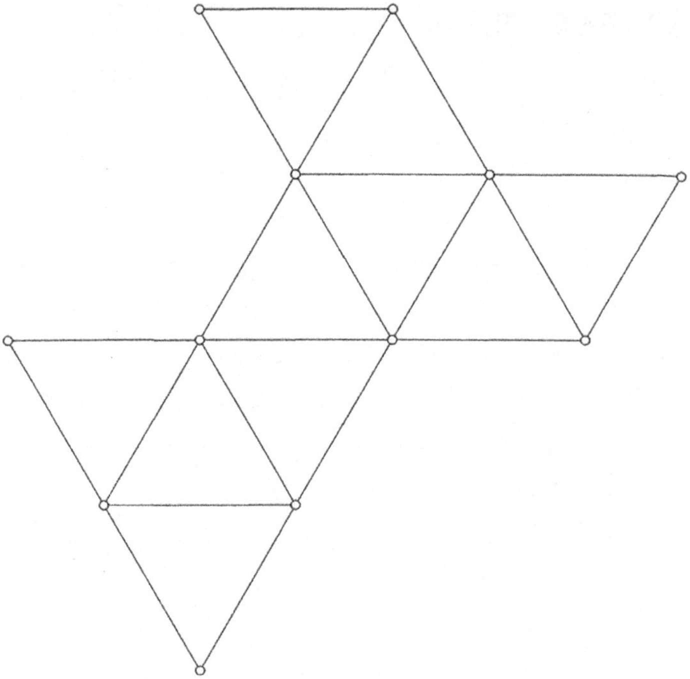

Puzzle 198: Base group $Z_{21}$

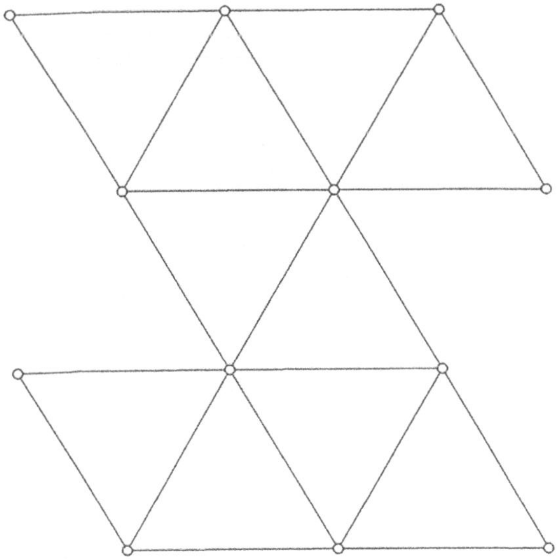

Puzzle 199: Base group $Z_{21}$

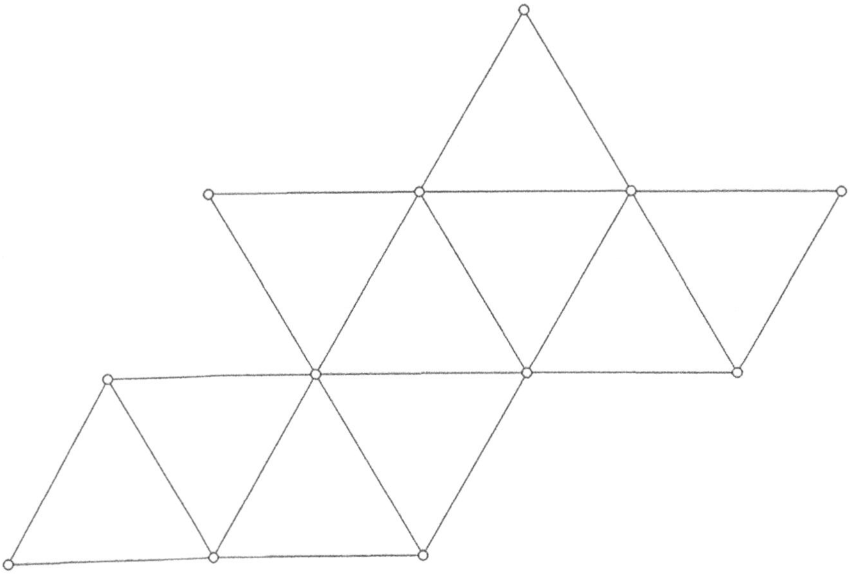

Puzzle 200: Base group $Z_{21}$

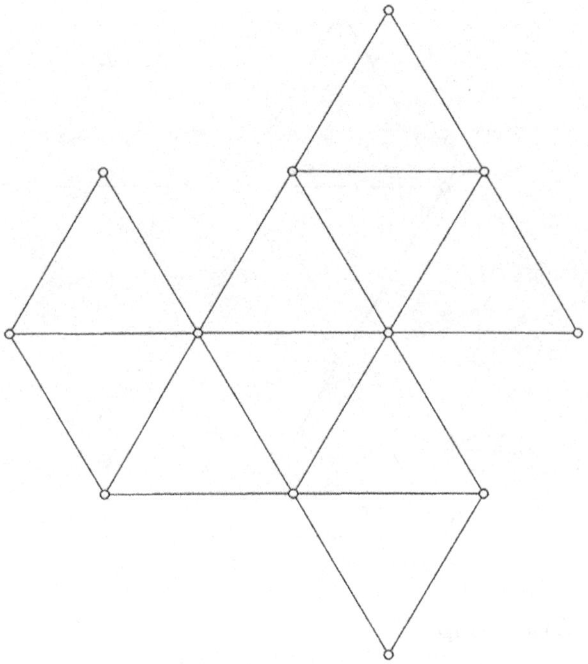

Puzzle 201: Base group $Z_{21}$

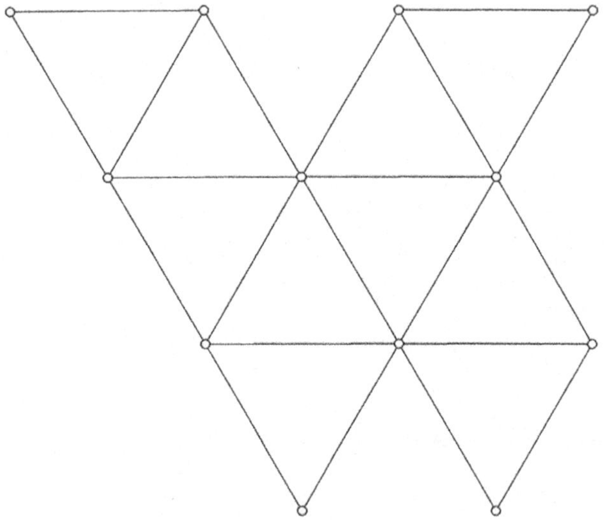

# 22. Wheels

(See also puzzles: 101, 103, 110, 122, 144)

Puzzle 202: Base group $Z_{18}$

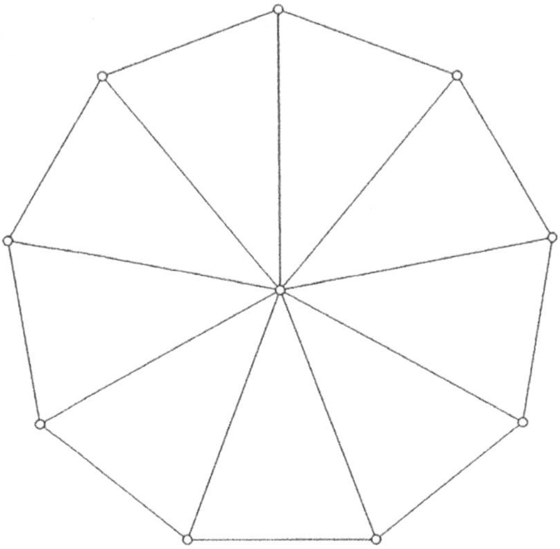

Puzzle 203: Base group $Z_{22}$

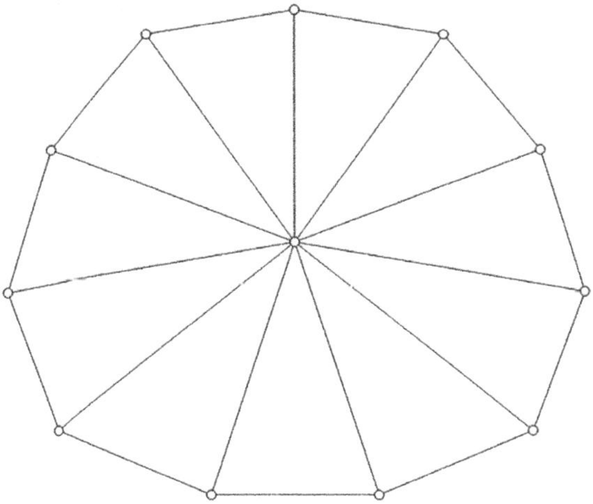

# 23. Doubles

Puzzle 204: Base group $Z_{20}$. Compare with puzzle 109.

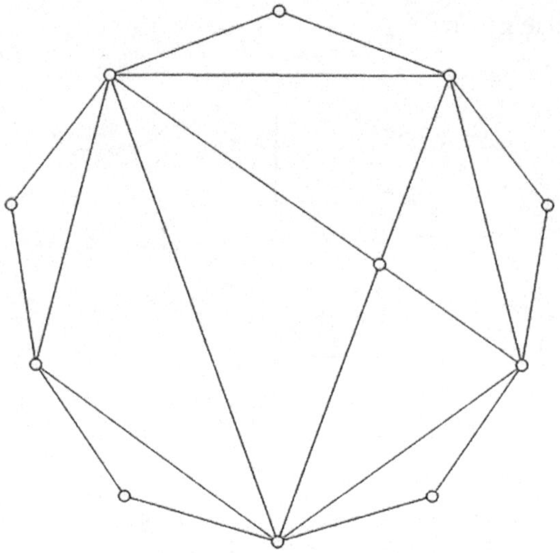

Puzzle 205: Base group $Z_{20}$, Compare with puzzle 110.

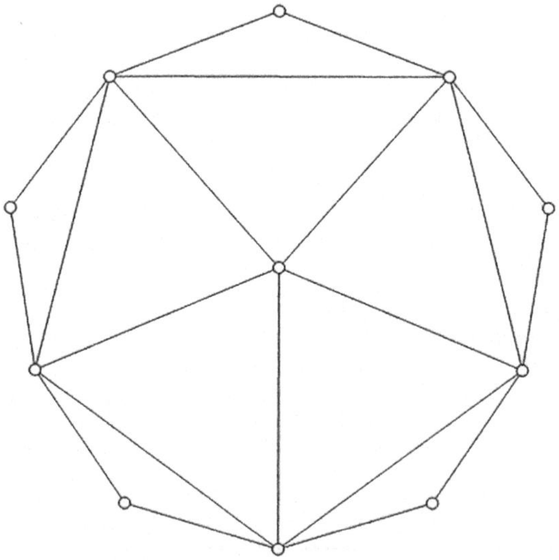

Puzzle 206: Base group $Z_{20}$. Compare with puzzle 111.

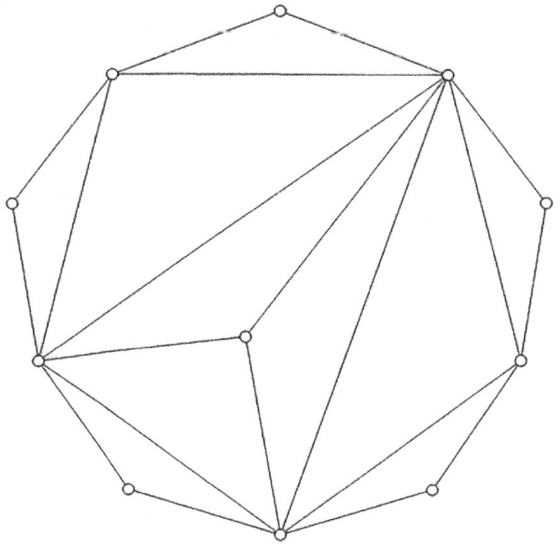

Puzzle 207: Base group $Z_{20}$. Compare with puzzle 112.

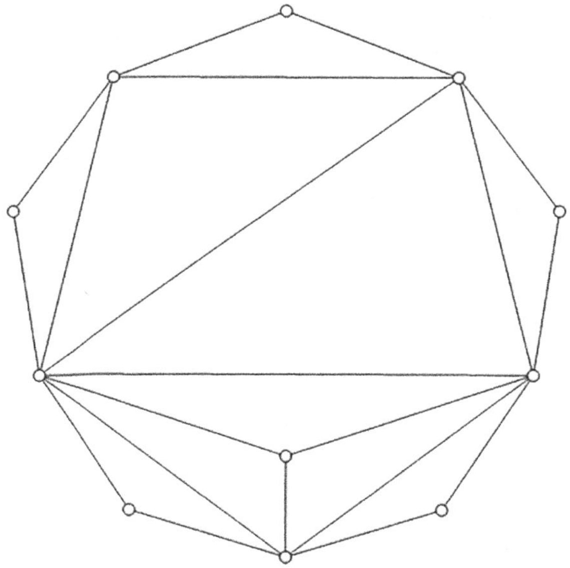

Puzzle 208: Base group $Z_{24}$. Compare with puzzle 119.

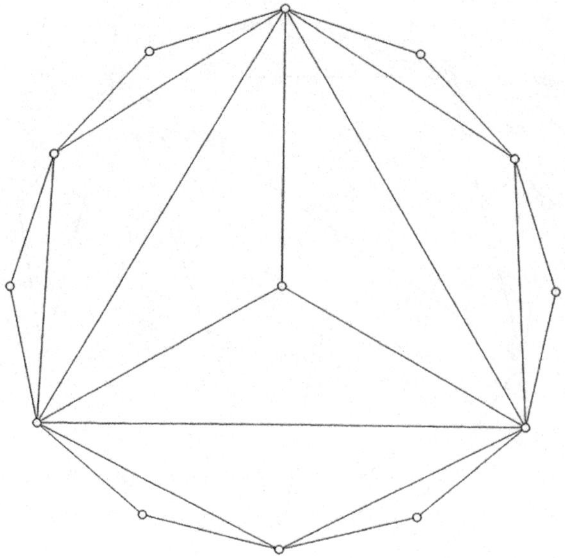

Puzzle 209: Base group $Z_{24}$. Compare with puzzle 120.

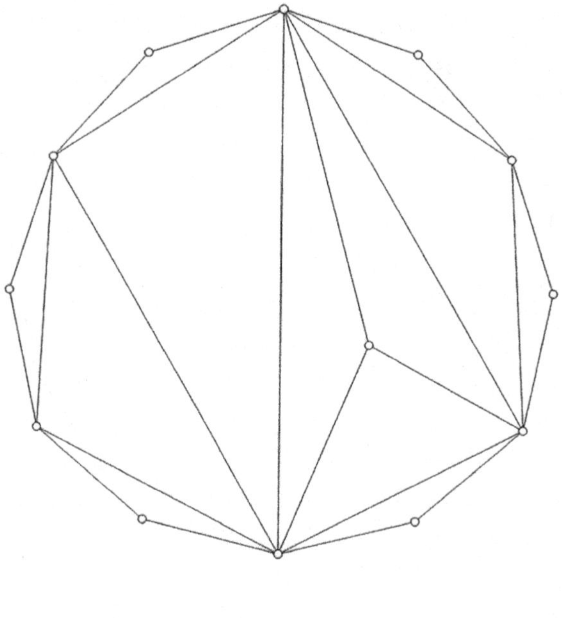

Puzzle 210: Base group $Z_{24}$. Compare with puzzle 121.

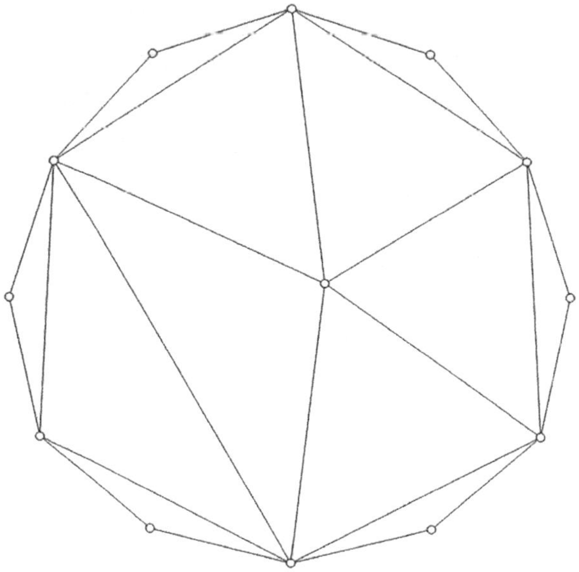

Puzzle 211: Base group $Z_{24}$. Compare with puzzle 122.

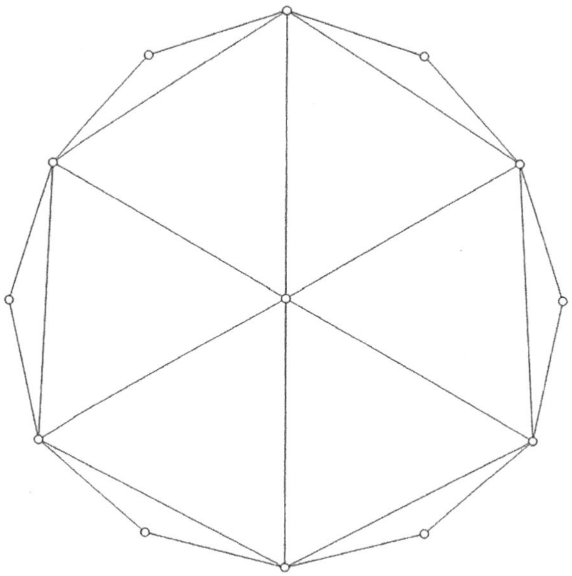

Puzzle 212: Base group $Z_{24}$. Compare with puzzle 123.

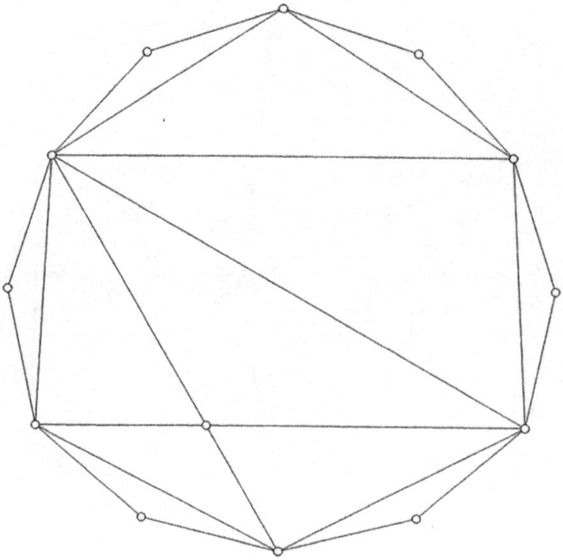

Puzzle 213: Base group $Z_{24}$. Compare with puzzle 124.

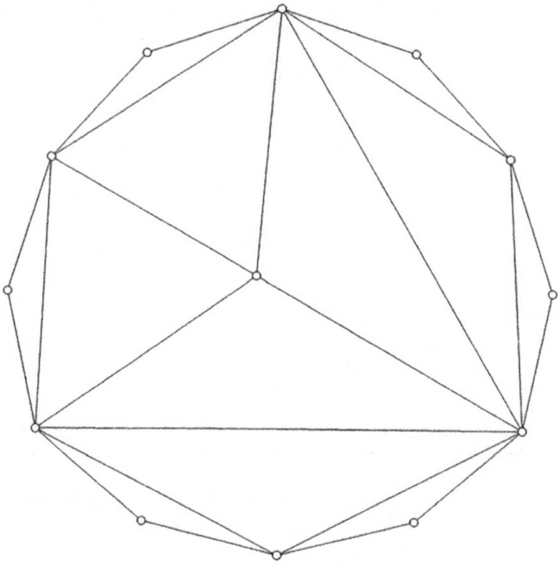

Puzzle 214: Base group $Z_{24}$ ; F = 0

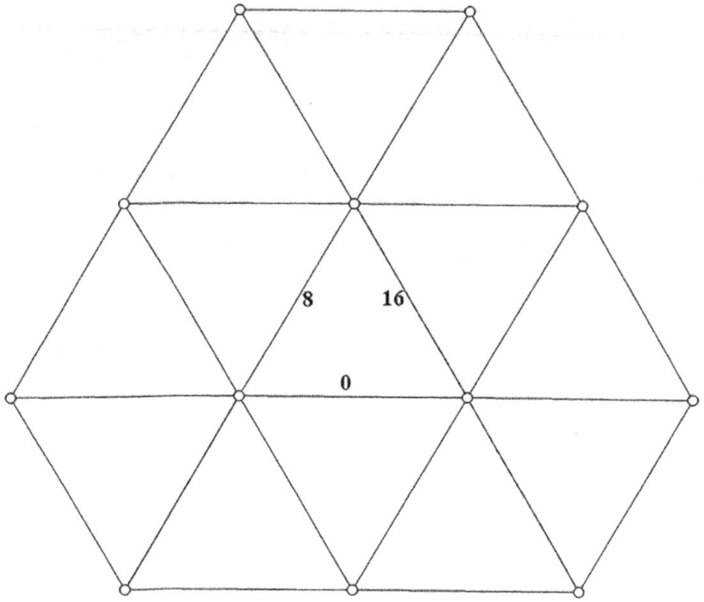

# Triangulations: problems and research questions

1. Find solutions to each of the following triangulations.

Puzzle 215: Base group $Z_{11}$

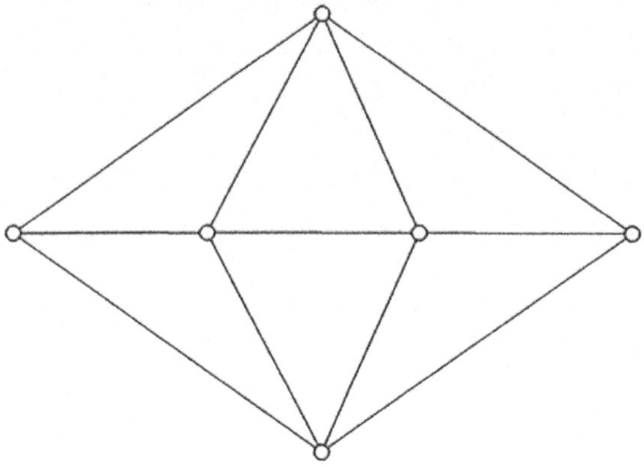

Puzzle 216: Base group $Z_{13}$

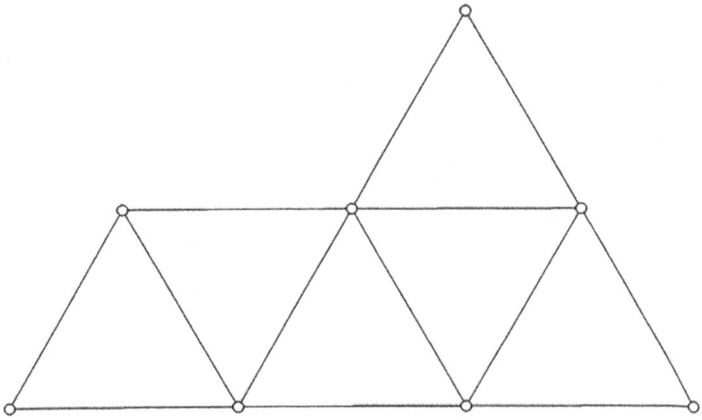

Puzzle 217: Base group $Z_{13}$

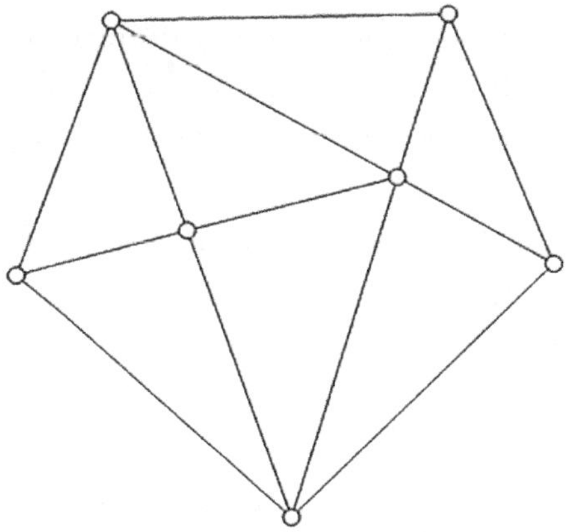

Puzzle 218: Base group $Z_{13}$

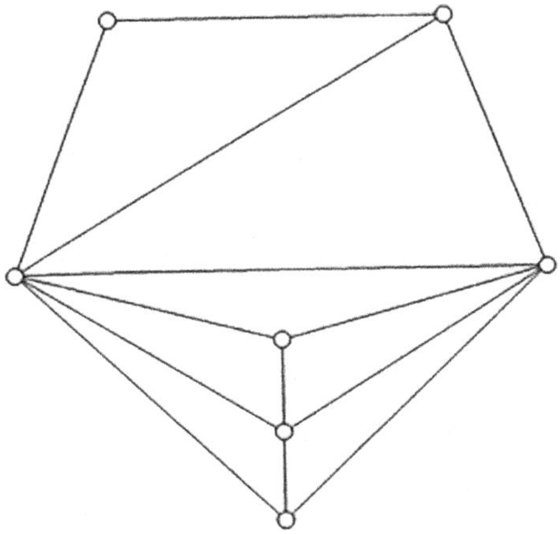

2. Let T be a polygonal triangulation with i interior edges; p interior points; b boundary edges and r interior triangles. Find a formula relating i, b and p and prove it.

3. With i, r, p defined as in (2), find a formula relating i, r and p and prove it.

4. Let T be a triangulation with a triangular boundary, (b = 3). Then T is called a *triangulation of the plane;* we call it simply a *pure triangulation.* The solution of a pure triangulation over $Z_n$ is *complete* in case the boundary triangle has the same edge sum s as each of the interior triangles. Show that s must be a multiple of 3.

5. * Puzzles 101, 126 are the only pure triangulations for which we have found complete solutions. Find more pure triangulations with complete solutions. Does every pure triangulation have a complete solution over $Z_n$?

6. A polygonal triangulation is *balanced* in case b = i. Use the formula for (2) to show that p = 1 for a balanced triangulation.

7. If T is a solved balanced polygonal triangulation with n edges and common triangle sum s then the *double T'* of T is obtained by doubling all the values assigned to the edges of T and then inserting the odd valued edges two at a time to each boundary edge of T to create a triangulation over $Z_{2n}$ with common triangle sum 2s. This is the method used to obtain the solutions given in Section 23. Use this method to construct doubles of balanced triangulations over $Z_{14}$ and $Z_{16}$.

8. * Given any balanced polygonal triangulation, is it always possible to construct its double? If so, then since the double of a balanced triangulation is also balanced, it would be possible to double and redouble ad infinitum any balanced triangulation.

9. In Section 20 and in previous Sections linear triangulations are given for which solutions have been obtained. We have included solutions over $Z_n$ for odd n up to n = 21. Show that for every odd n, the linear triangulation with n edges has a monotone solution over $Z_n$. Note that this is equivalent to constructing a sequence of length n consisting of the full set of elements of $Z_n$ such that every other triple of consecutive elements has the same sum (mod n).

10. Show that every wheel with an even number m of vertices can be solved over $Z_{2(m-1)}$. (A *wheel* is a planc graph consisting of a polygon and a central point which is joined by an edge to each vertex of the polygon.)

11. * Puzzle 103 with m = 5 vertices, and puzzle 122 with m = 7 vertices are the only wheels for which solutions have been found over $Z_{2(m-1)}$ with m odd. Find solutions for other wheels with an odd number m of vertices.

12. * Let $D_m$ denote the plane graph consisting of a path P with m-2 points and two additional points x, y which are joined by edges to each point of P. Puzzles 103, 215 represent $D_5$, $D_6$. The graph $D_7$ with solution over $Z_{14}$ is given below. Describe a procedure for solving $D_m$ for m>7.

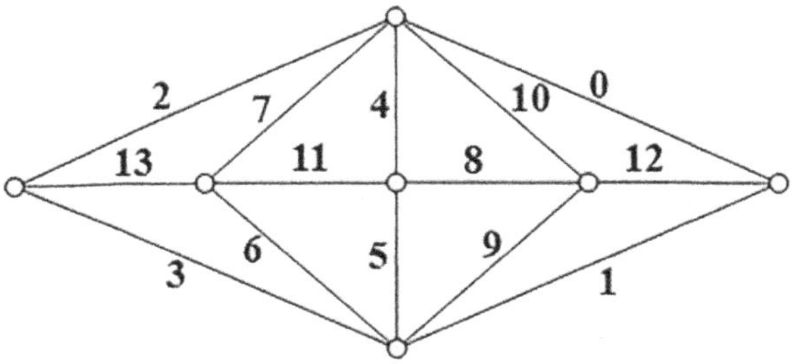

# SECTION III

# POLYHEXES

These puzzles consist of regular hexagons which are fitted edge to edge. We call the resulting figure a *polyhex*. By affixing a point to the center of each component hexagon and then including an edge joining the centers of each pair of adjacent hexagons, we obtain the *adjacency graph* of the polyhex. The adjacency graph for each polyhex is indicated using dashed lines. The adjacency graph is a simple structure from which the corresponding polyhex can be constructed. All solutions to the puzzles in this Section are given using the inverse notation described in the Introduction. This is the notation we recommend for solving polyhex puzzles.

# 24. Tree Polyhexes

In mathematics a *graph* G = (V,E) is a set V of points (or vertices) and a set E consisting of pairs of points from V. The elements of E are called edges and are depicted by a line joining the pair of points involved. For example, in puzzle 219 the vertex set V contains 14 points and the edge set E contains 216 edges. (We are not counting the small points and dashed edges). A *path* in a graph is a sequence $\langle e_1, e_2, \ldots, e_t \rangle$ of edges such that each edge shares a vertex only with its predecessor. A *cycle* or *closed path* is a path in which the last vertex is connected by an edge to the first vertex. A *tree* is a connected graph with no cycles. In this Chapter the adjacency graph for each polyhex is a tree.

Puzzle 219: Base group $Z_{14}$ = {0, 1,-1,2,-2,3,-3,4,-4,5,-5,6,-6,7}

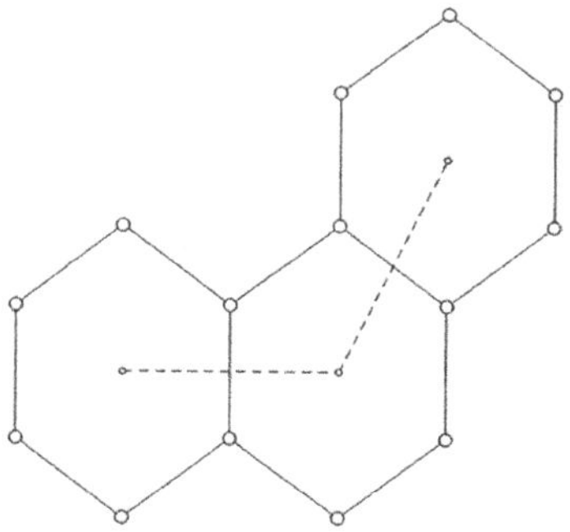

Puzzle 220: Base group $Z_{18}$ = {0, 1,-1,2,-2, …, 8,-8,9}

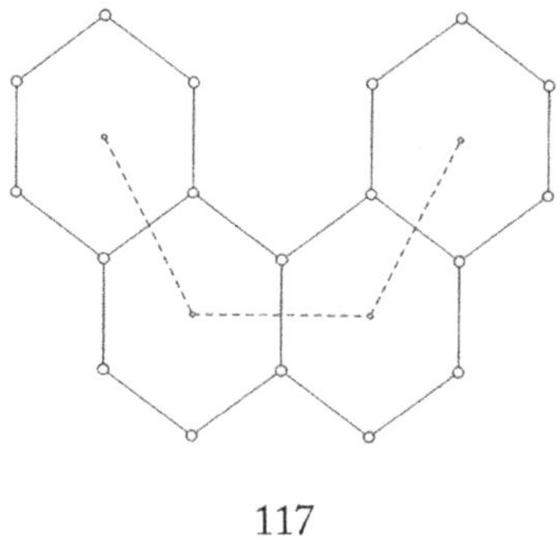

Puzzle 221: Base group $Z_{18}$ = {0, 1,-1,2,-2, …, 8,-8,9}

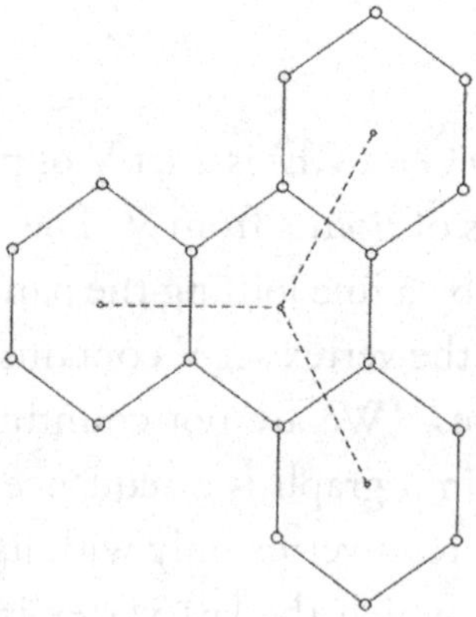

Puzzle 222: Base group $Z_{30}$ = {0, 1,-1, 2,-2, …, 14, -14, 15}

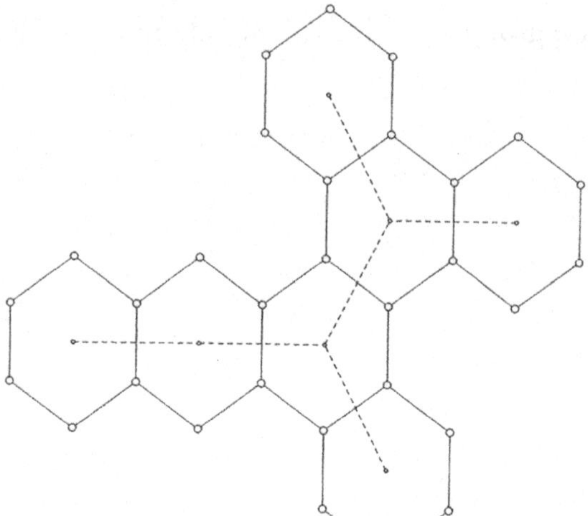

# 25. 3,4 and 5-hexes

An *n-hex* is a polyhex consisting of n hexagonal components. In this Chapter the polyhexes have three four or five component hexagons and none have adjacency graphs which are trees.

Puzzle 223: (the *3-hex triangle*) Base group $Z_{13} = \{0, 1,-1,2,-2, \ldots, 6,-6\}$

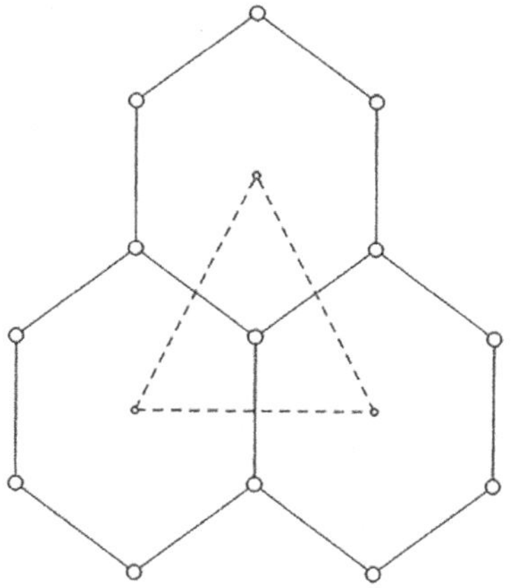

Puzzle 224: Base group $Z_{17} = \{0, 1,-1,2,-2, \ldots, 8,-8\}$

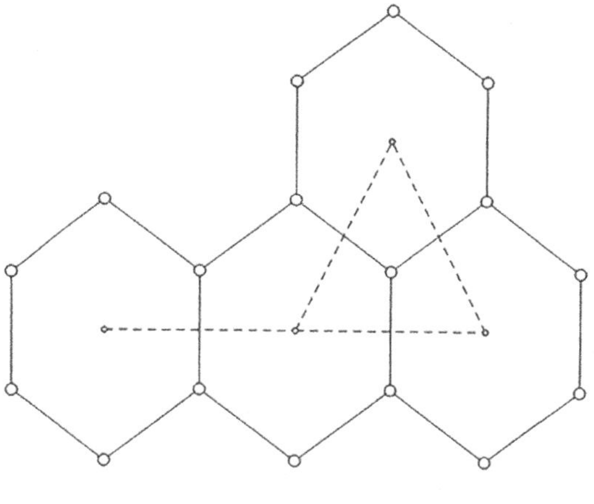

Puzzle 225: (the *4-hex diamond*) Base group $Z_{16}$

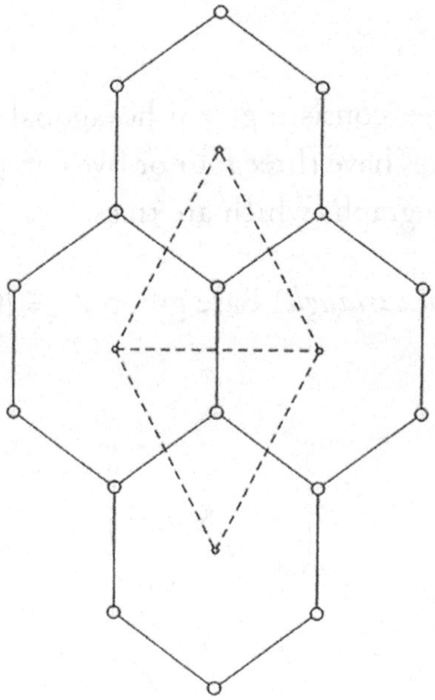

Puzzle 226: Base group $Z_{20}$ = {0,1,-1,2,-2, …, 9,-9, 10}

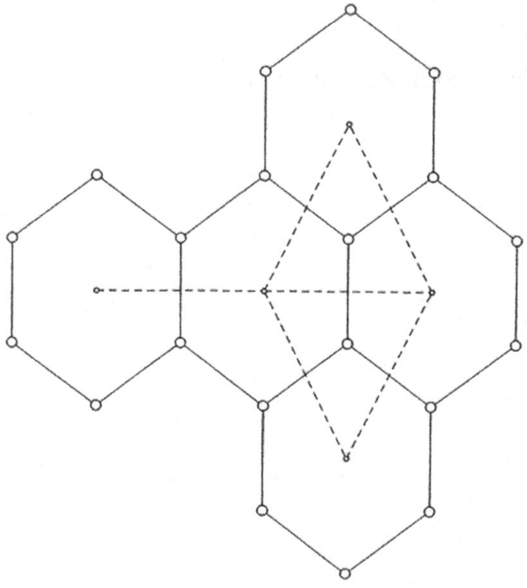

Puzzle 227: Base group $Z_{21} = \{0, 1,-1,2,-2, \ldots, 10, -10\}$

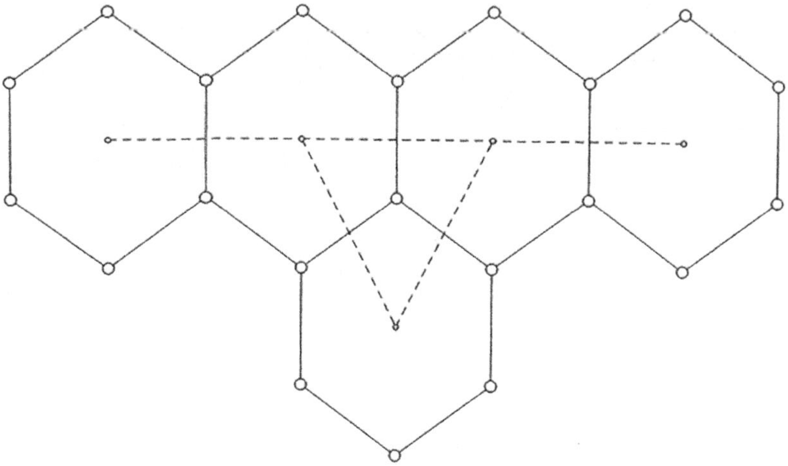

Puzzle 228: Base group $Z_{21}$

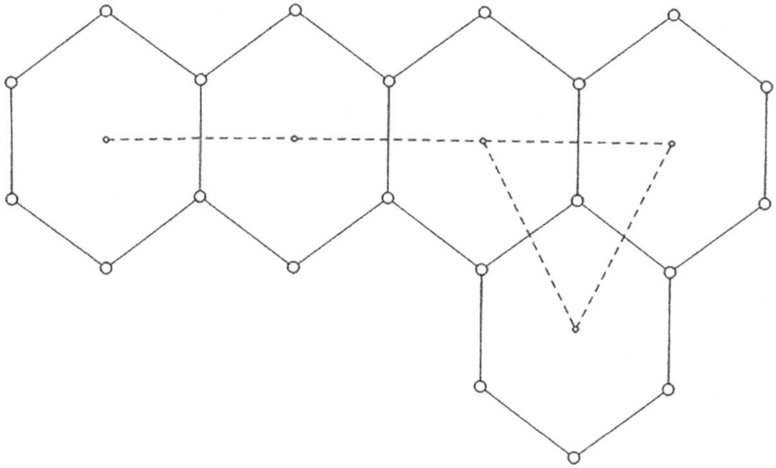

Puzzle 229: Base group $Z_{19}$ = {0,1,-1,2,-2, …, 9,-9}

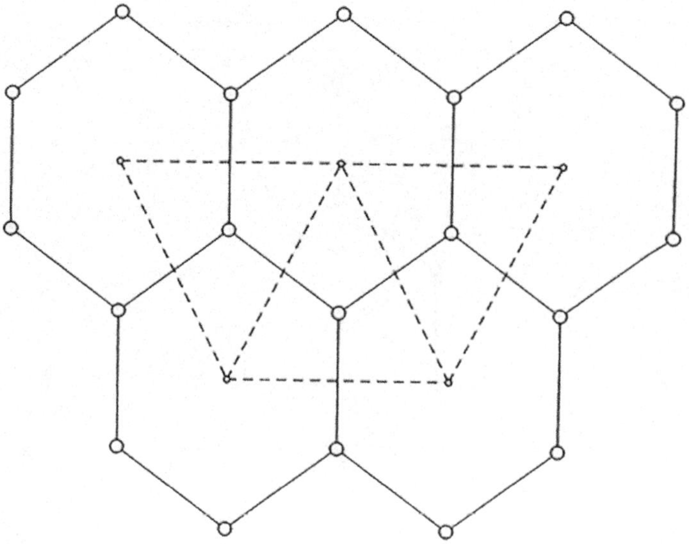

Puzzle 230: Base group $Z_{20}$ = {0,1,-1,2,-2, …, 9,-9,10}

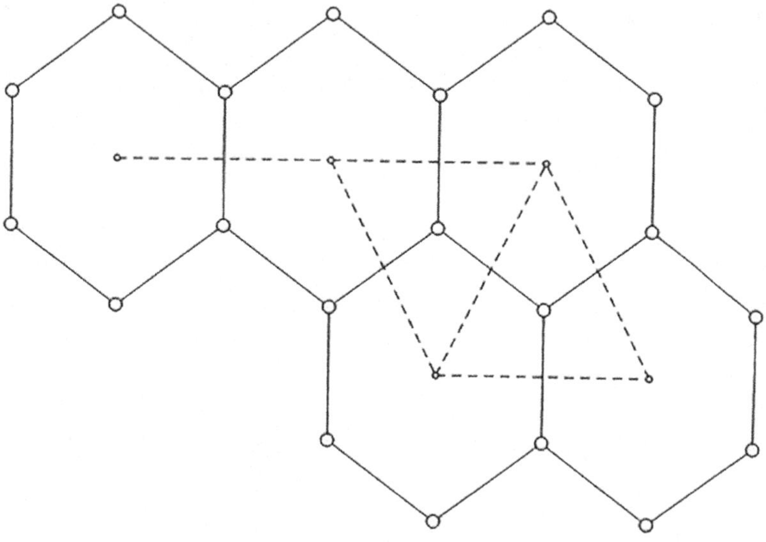

# 26. 6, 7 and 8-hexes

Puzzle 231: Base group $Z_{24} = \{0,1,-1,2,-2, \ldots, 11,-11,12\}$

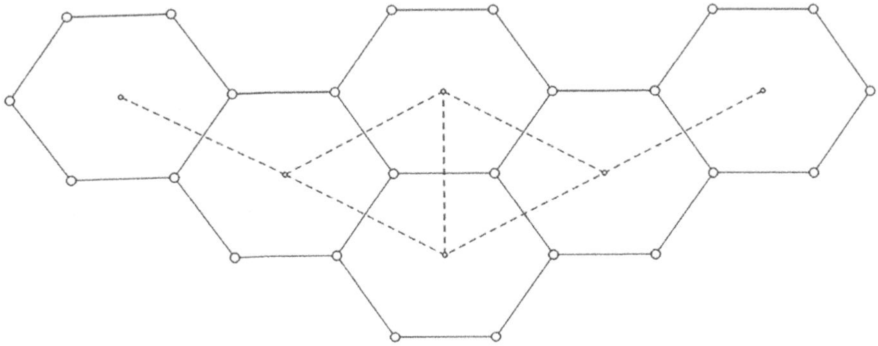

Puzzle 232: Base group $Z_{24}$

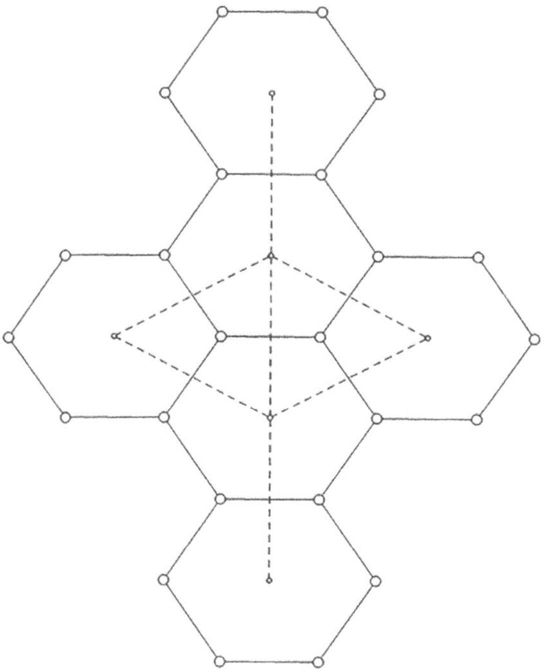

Puzzle 233: Base group $Z_{24}$ = {0,1,-1,2,-2, …, 11,-11,12}

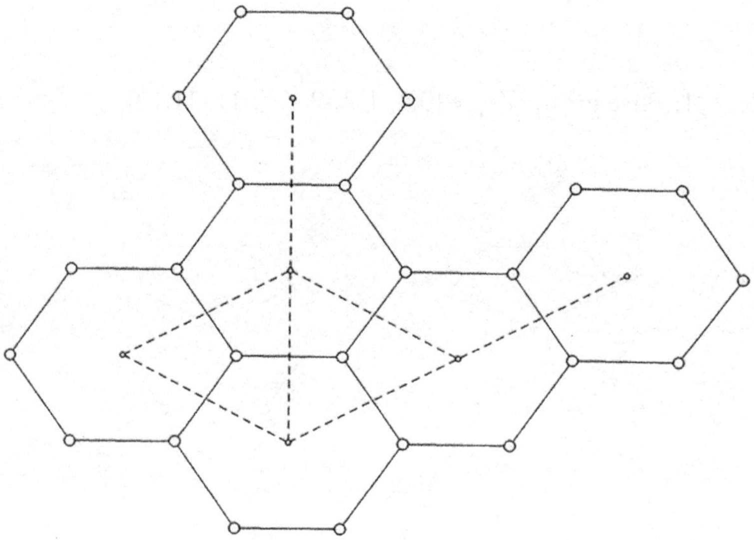

Puzzle 234: Base group $Z_{25}$ = {0,1,-1,2,-2, …, 12,-12}

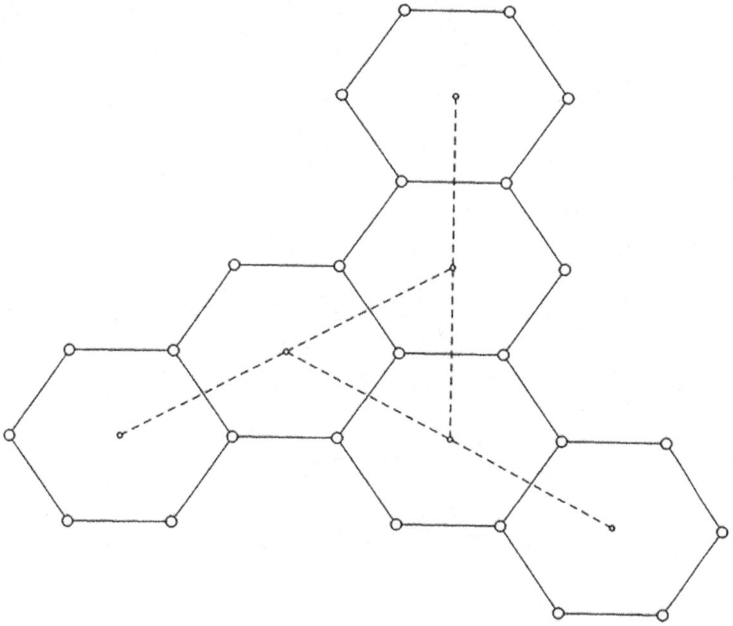

Puzzle 235: Base group $Z_{22} = \{0,1,-1,2,-2, \ldots, 10,-10,11\}$

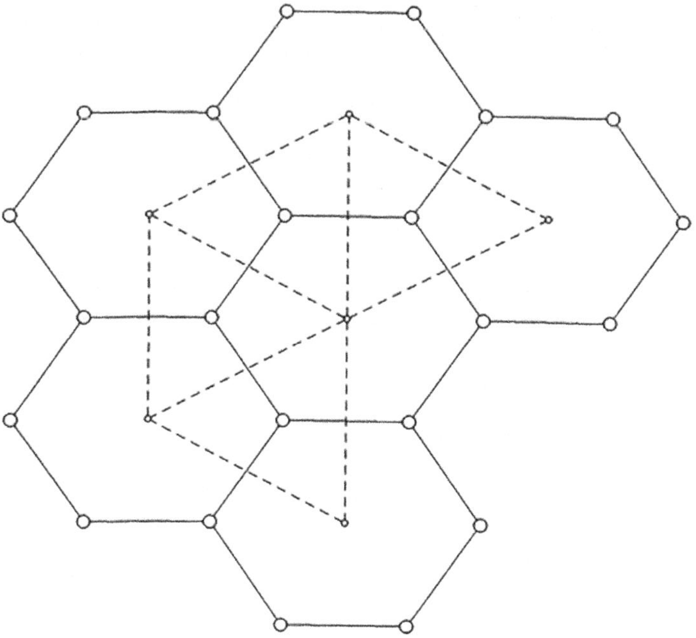

Puzzle 236: Base group $Z_{22} = \{0,1,-1,2,-2, \ldots, 10,-10,11\}$

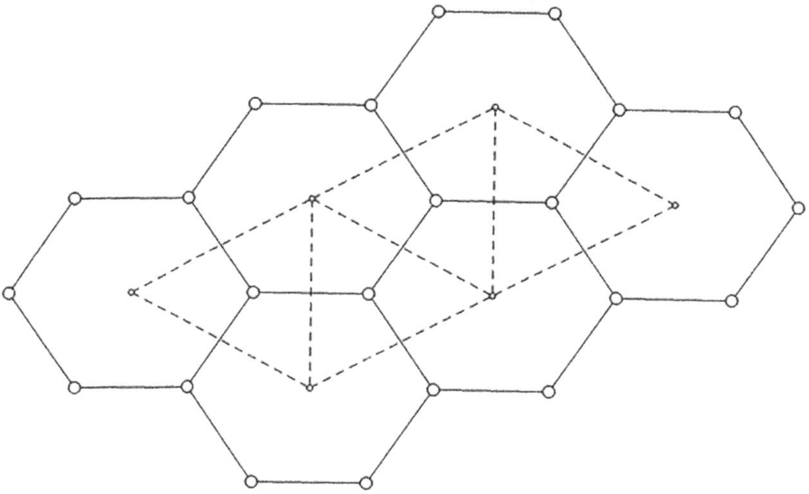

Puzzle 237: *(6-hex triangle)* Base group $Z_{22}$ = {0,1,-1,2,-2, …, 10,-10,11}

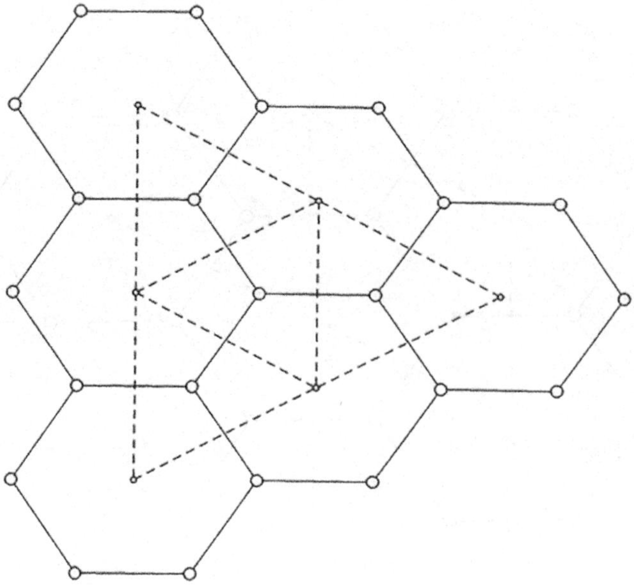

Puzzle 238: (*6-hex hexagon)* Base group $Z_{24}$ = {0,1,-1,2,-2, …, 11,-11,12}

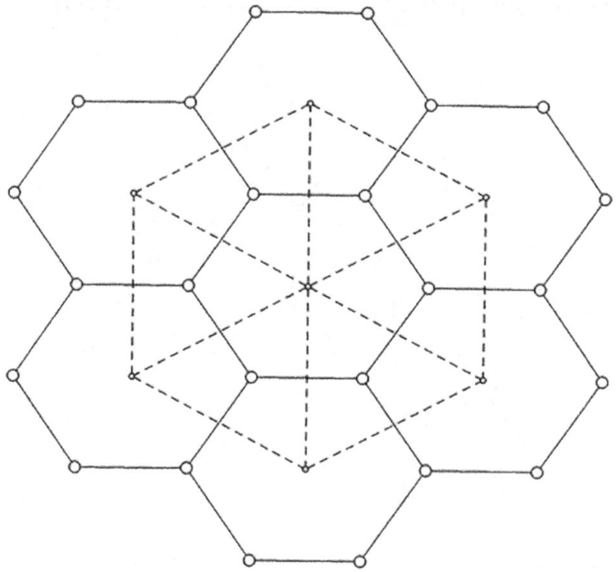

Puzzle 239: Base group $Z_{25} = \{0,1,-1,2,-2, \ldots, 12,-12\}$

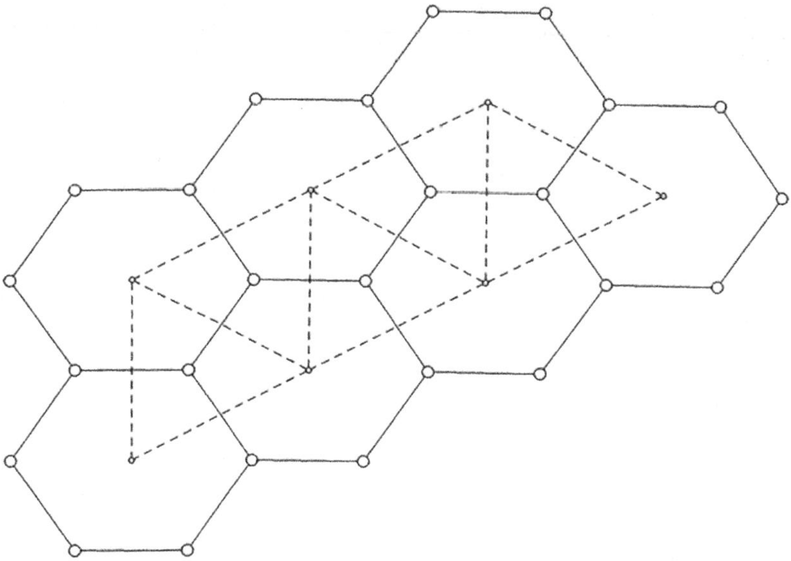

Puzzle 240: Base group $Z_{24} = \{0,1,-1,2,-2, \ldots, 11,-11,12\}$

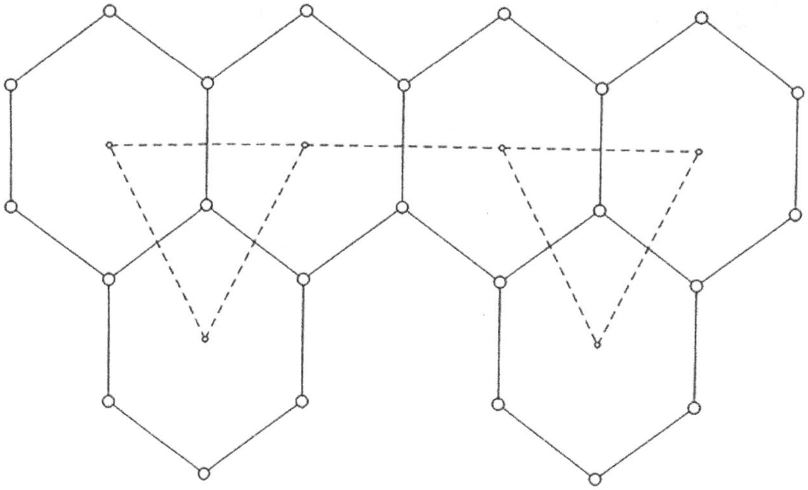

Puzzle 241: Base group $Z_{27}$ = {0,1,-1,2,-2, …, 13,-13}

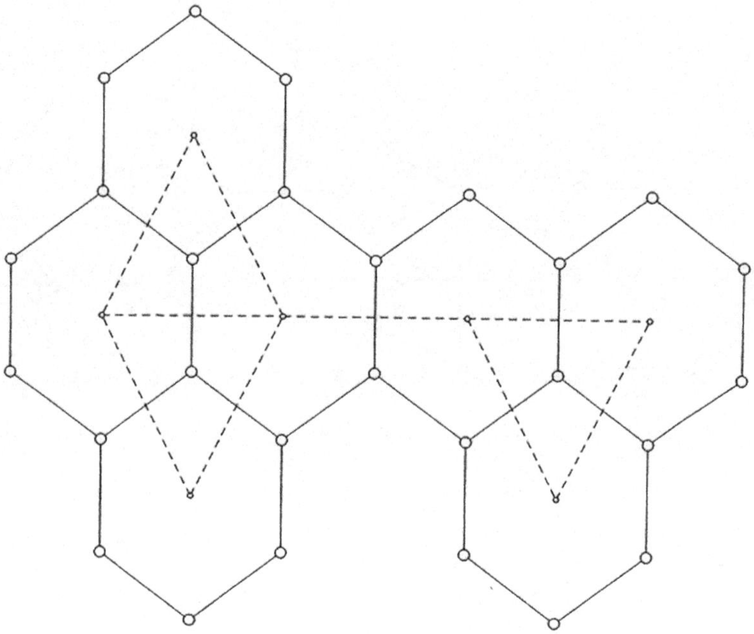

Puzzle 242: Base group $Z_{25}$ = {0,1,-1,2,-2, …, 12,-12}

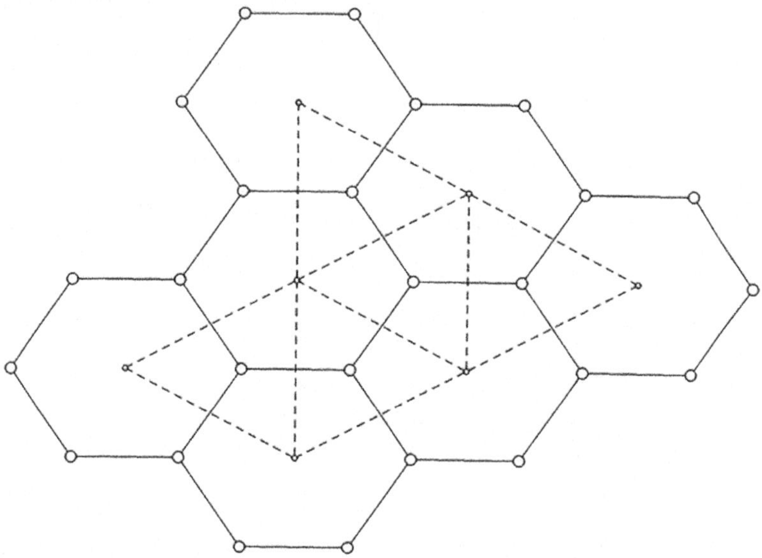

Puzzle 243: Base group $Z_{25} = \{0,1,-1,2,-2, \ldots, 12,-12\}$

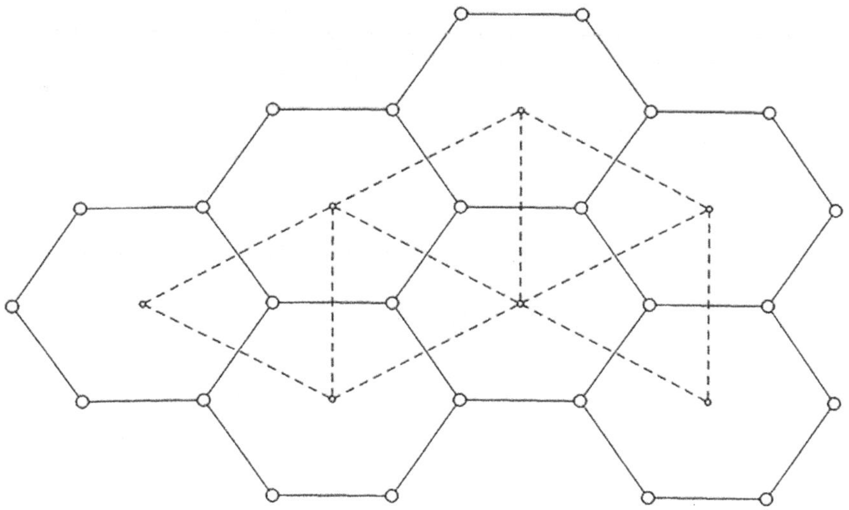

Puzzle 244: Base group $Z_{27} = \{0,1,-1,2,-2, \ldots, 13,-13\}$

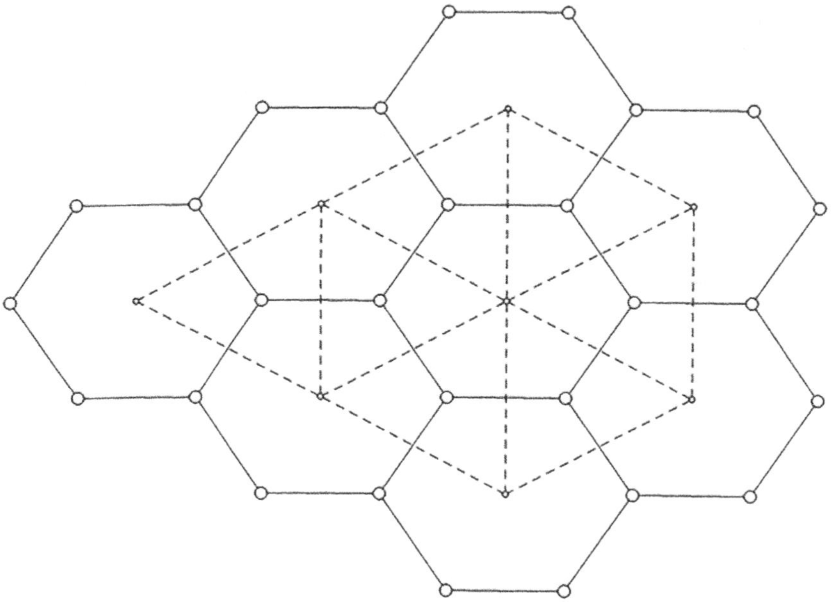

Puzzle 245: Base group $Z_{28} = \{0,1,-1,2,-2, \ldots, 13,-13,14\}$

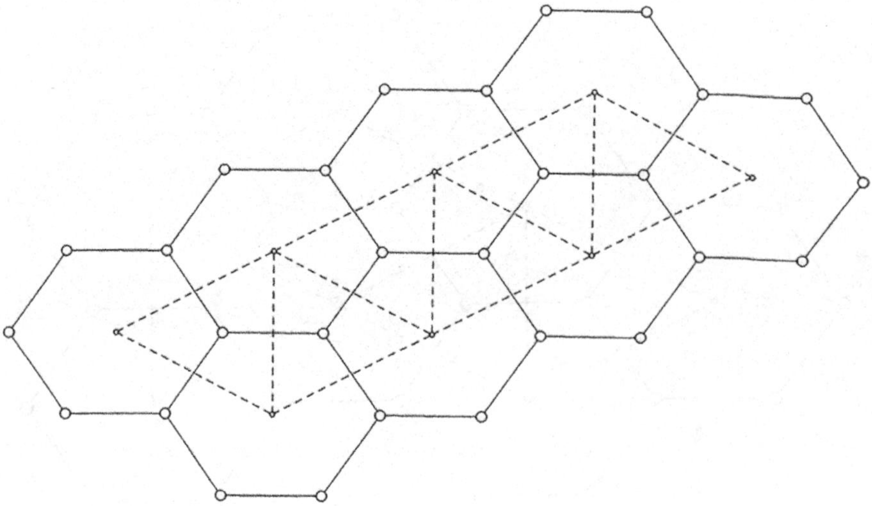

Puzzle 246: Base group $Z_{30} = \{0,1,-1,2,-2, \ldots, 14,-14,15\}$

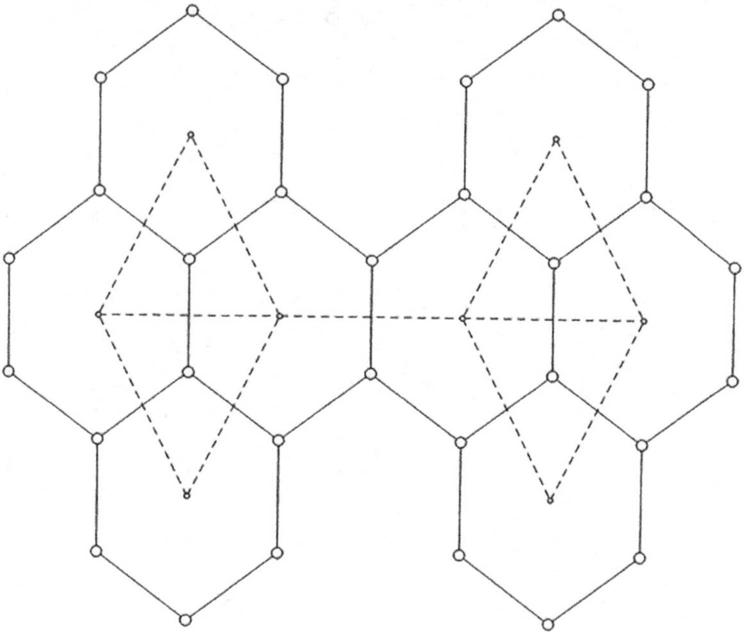

# 27. Extensions

Polyhex solutions are often obtained by extending the solution to a smaller polyhex. If the smaller polyhex has m vertices, we call it a $Z_m$ - *inclusion.* A solved polyhex P may have any number of inclusions which must be *nested.* For example, if P contains a $Z_{16}$ inclusion with vertices labeled {0,1,-1,2,-2, …, 7,-7,8}, then a larger inclusion, say a $Z_{24}$ - inclusion, must contain these 16 points in addition to the points {-8,9,-9,10,-10, 11, -11,12}. To make the puzzle solutions in this Section more accessible the inclusions are indicated by the following numbering system. Suppose the inclusions are nested in the order $I_1 \subset I_2 \subset I_3 \subset \ldots$ P. The number 1 will be placed in the center of each hexagon belonging to $I_1$, The numbers 1 and 2 will be placed in the centers of hexagons belonging to $I_2$. In general the inclusion $I_m$ will consist of all hexagons of P whose center numbers are less than or equal to m. Consider the example below.

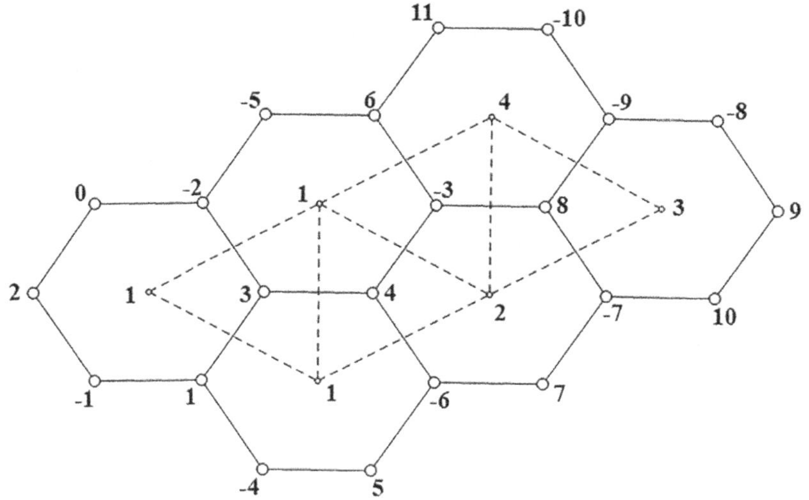


In this example $I_1$ is a $Z_{13}$ inclusion; $I_2$ is a $Z_{16}$ inclusion; $I_3$ is a $Z_{20}$ inclusion and $I_4$ = P is the entire polyhex with a solution over $Z_{22}$.

Puzzle 247: Base group $Z_{40}$

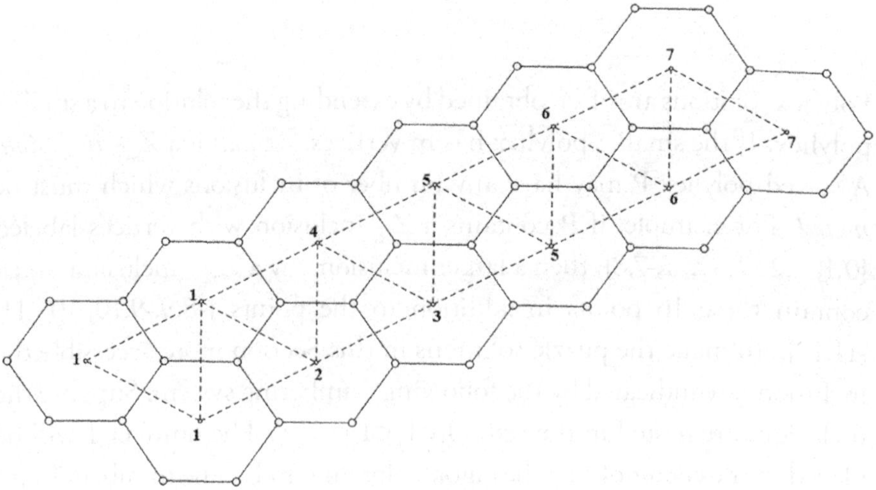


Puzzle 248: Base group $Z_{36}$

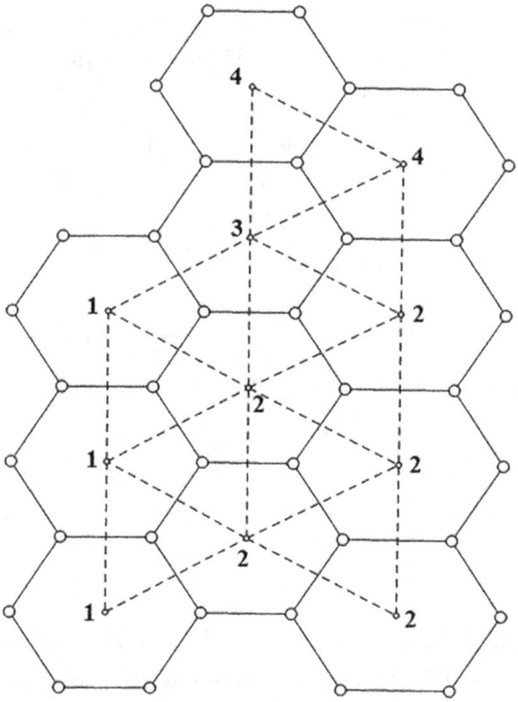

Puzzle 249: Base group $Z_{38} = \{0,1,-1,2,-2, \ldots, 18,-18,19\}$

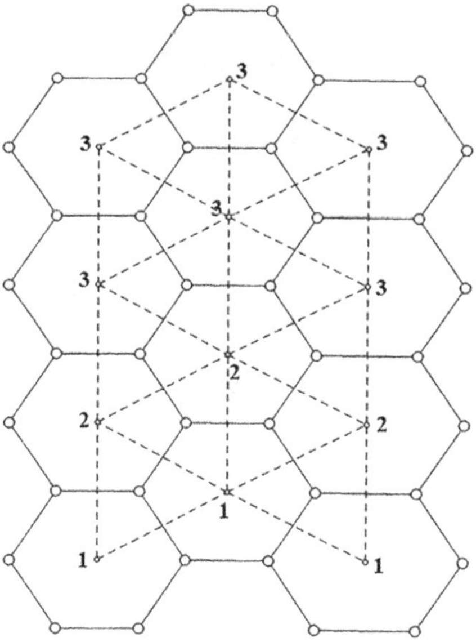


Puzzle 250: Base group $Z_{40} = \{0,1,-1,2,-2, \ldots, 19,-19,20\}$

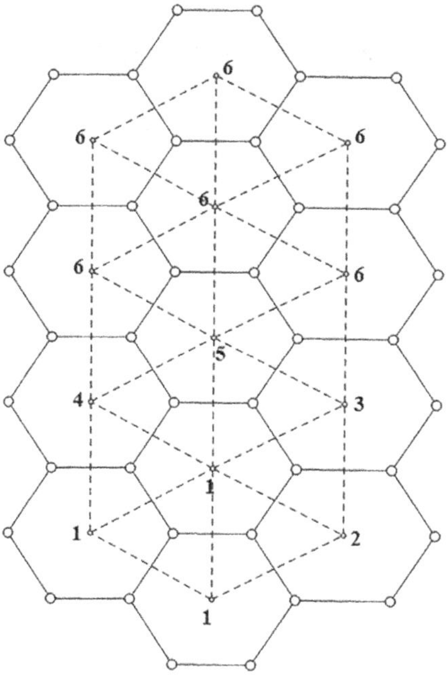

Puzzle 251: Base group $Z_{36}$ = {0,1,-1,2,-2, …, 17,-17,18}

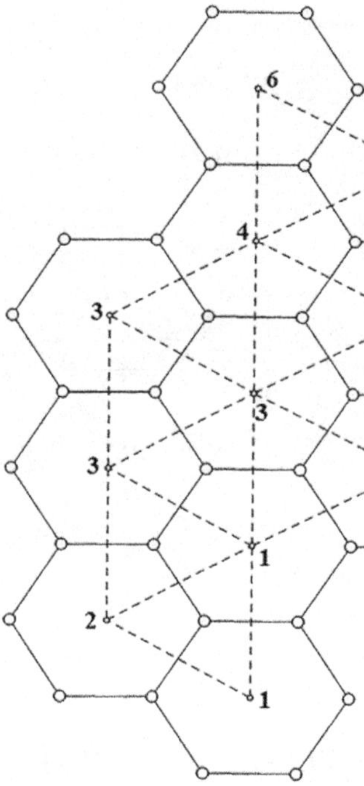


Puzzle 252:(12-hex parallelogram) Base group $Z_{38}$ = {0,1,-1,2,-2, …, 18,-18,19}

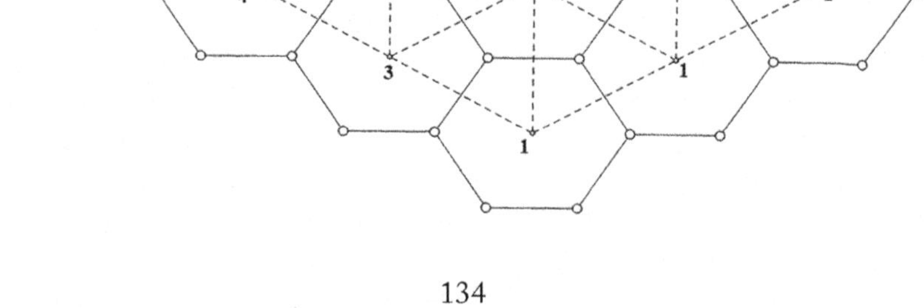

Puzzle 253: 10-hex triangle with base group $Z_{33}$ = {0,1,-1,2,-2, …, 16,-16}

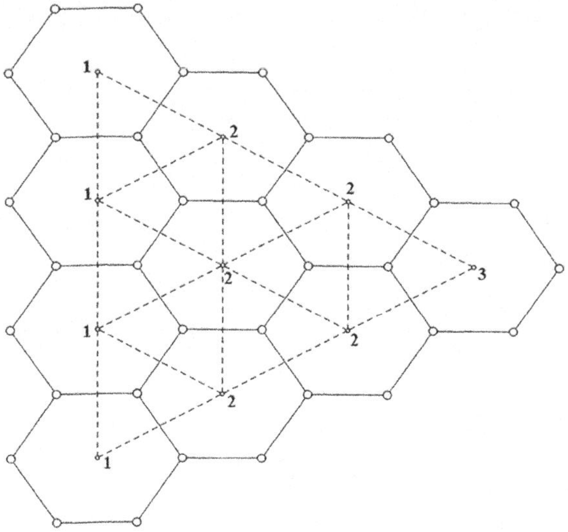


Puzzle 254: 9-hex rhombus with base group $Z_{28}$ ={0,1,-1,2,-2, …, 13,-13,14}

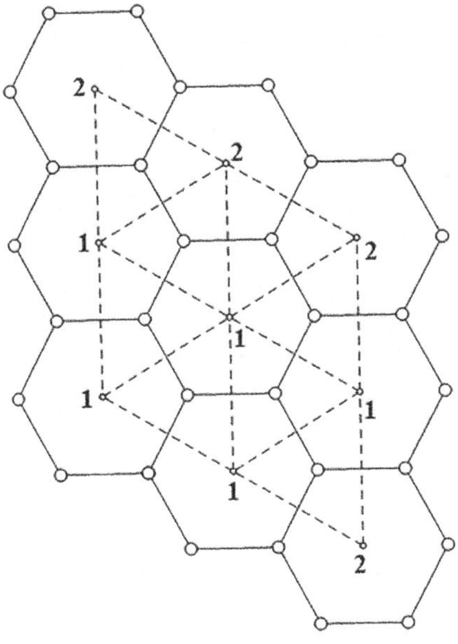

Puzzle 255: (12-hex trapezoid) Base group $Z_{38}$ = {0,1,0-1,2,-2, …, 18,-18,19}

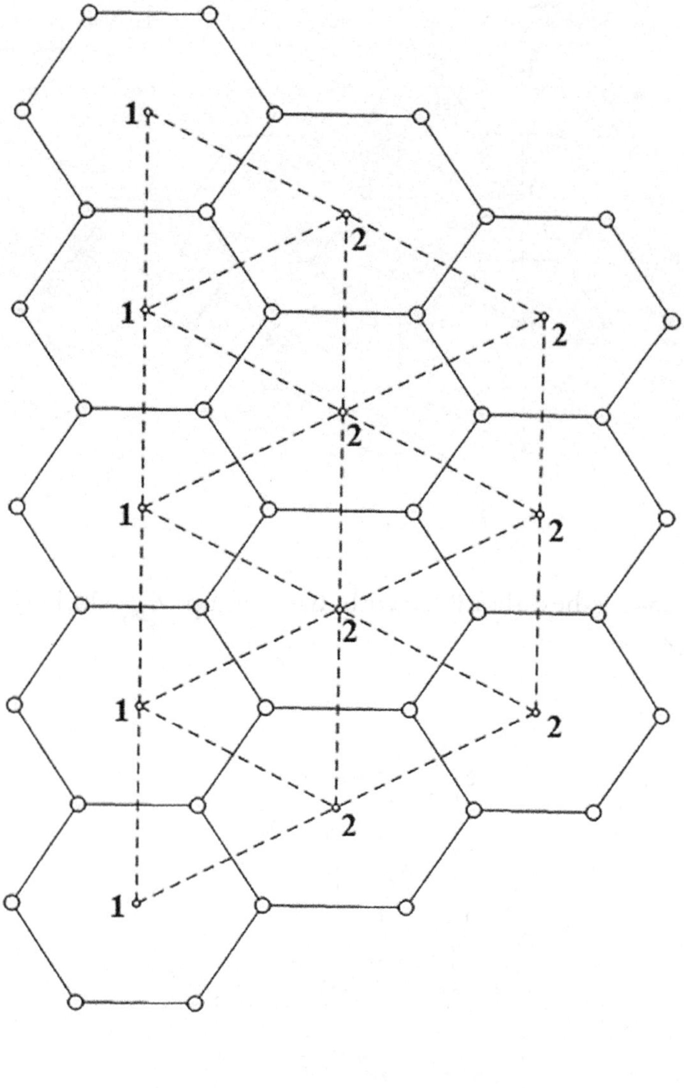

Puzzle 256: 15-hex trapezoid with base group $Z_{46}$ = {0,1,-1,2,-2,..., 22,-22,23} with extension to 21-hex triangle with base group $Z_{61}$ = {0,1,-1,2, -2, ...,30, -30}

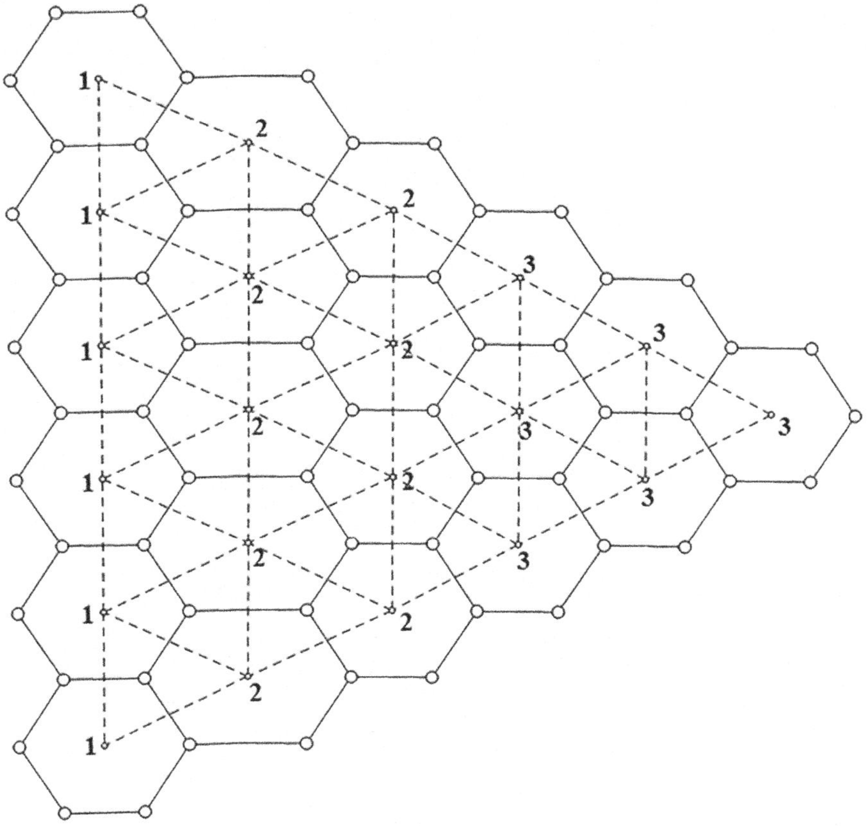

Puzzle 257:(15-hex parallelogram)Base group $Z_{46}$ = {0,1,-1,2,-2, …, 22,-22,23)

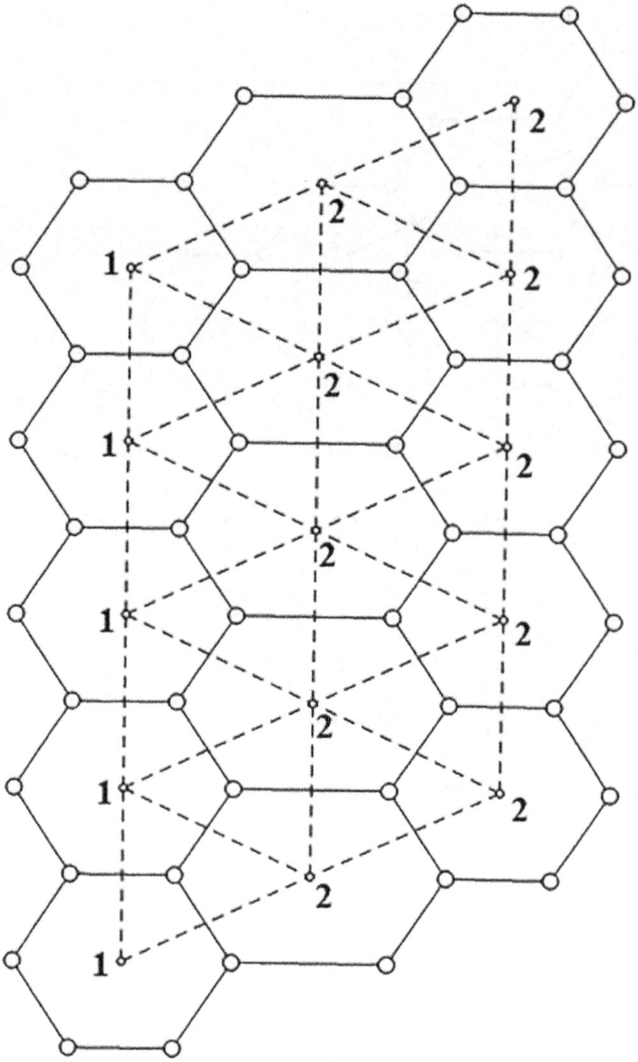

# Polyhexes: problems and research questions

1. Show that every polyhex with n points and whose adjacency graph is a tree, has a solution over $Z_n$.

2. A *bridge* is an edge of a graph whose removal disconnects the graph. Every edge of a tree is a bridge, thus the adjacency graphs of the polyhexes in Chapter 24 are composed entirely of bridges. Bridges also occur in the adjacency graphs of puzzles 224, 226, 227, 228, 230-234,240, 241, 246. Solve each of the polyhexes whose adjacency graphs are given below. Each graph contains several bridges. Extensions are indicated in the usual way by labeling the points of the adjacency graphs.

(i) Base group $Z_{44}$

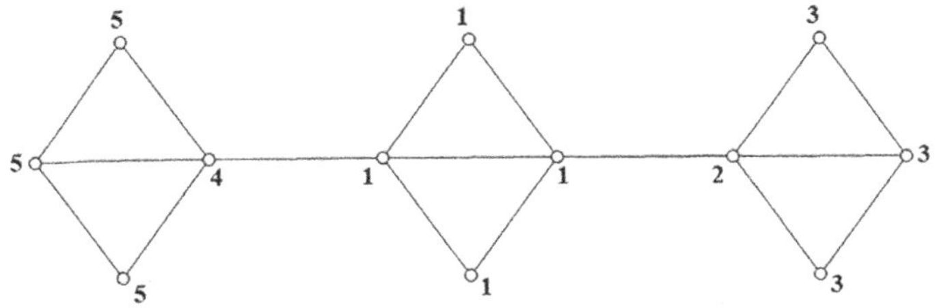

(ii) Base group $Z_{46}$

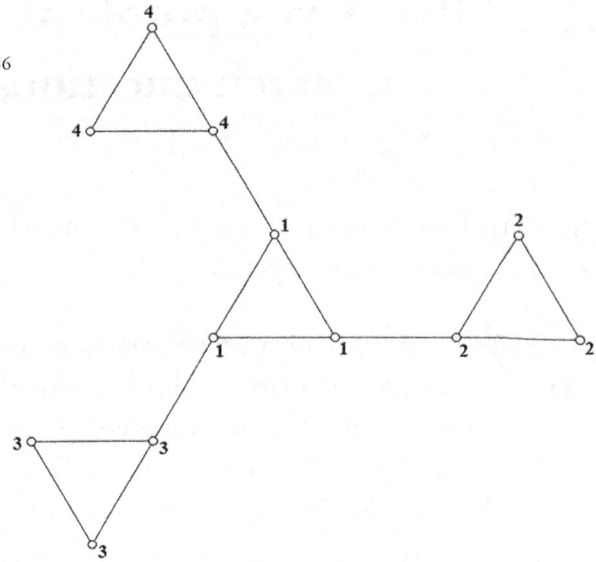


(iii) Base group $Z_{48}$

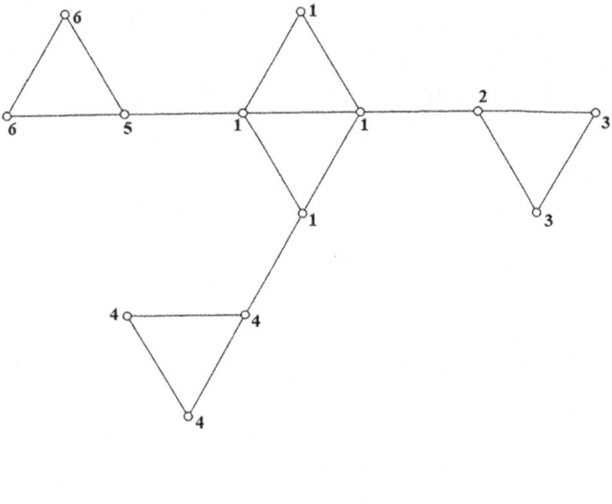


3. Polyhexes 225, 236, 245 and 247 all have adjacency graphs which are linear equilateral triangulations with an even number n of points. Show that every such polyhex has a solution over $Z_n$.

4. * Polyhexes 223, 229, 239 all have adjacency graphs which are linear equilateral triangulations with an <u>odd</u> number n of points. Can these be solved over $Z_n$ for every odd n?

5. * Puzzles 223, 237, 253(as inclusion) and 256 are *triangle polyhexes..* Puzzles 229, 254(as inclusion), 257, 258 are *polyhex parallelograms.* Puzzles 229, 239, 255, 256 (as inclusion) are *polyhex trapezoids.* Puzzles 238, 250 are *polyhex hexagons.* Find a method for solving larger polyhexes in each of these categories.

6. * Polyhex puzzles can also be constructed in which the edges instead of the vertices are labeled with elements from $Z_n$. In the example below there are 16 vertices but 19 edges, so elements from $Z_{19}$ are used to provide the indicated solution. Try solving the polyhex puzzles given in this Section using this edge labeling.

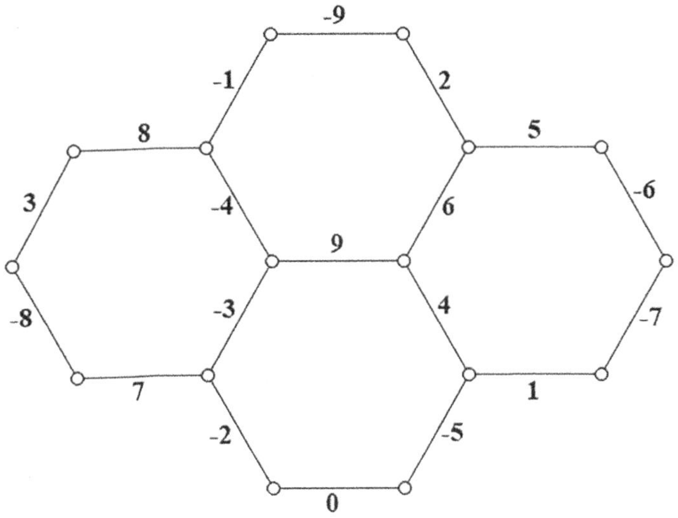

# Polycube Puzzle Solutions

Puzzle 1

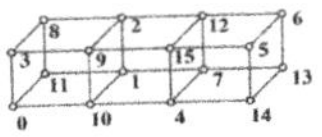

Puzzle 2

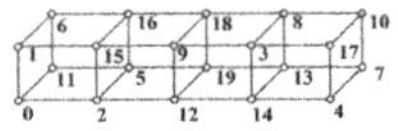

Puzzle 3

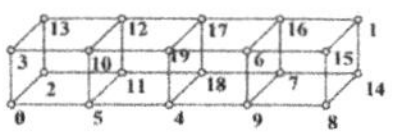

Puzzle 4

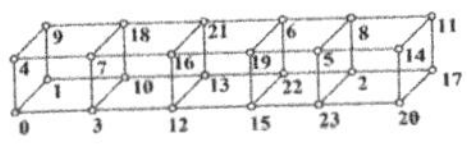

Puzzle 5

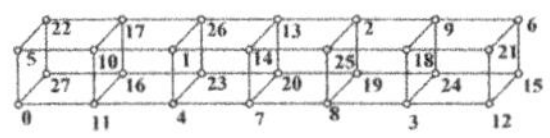

Puzzle 6

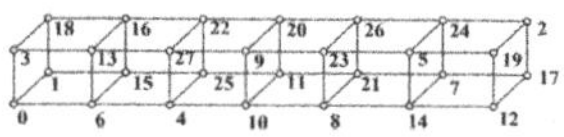

Puzzle 7

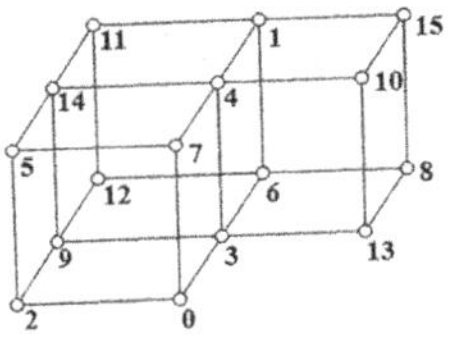


Puzzle 8

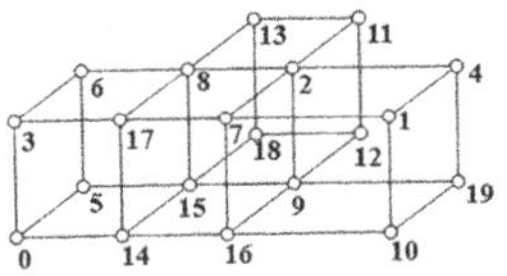


Puzzle 9

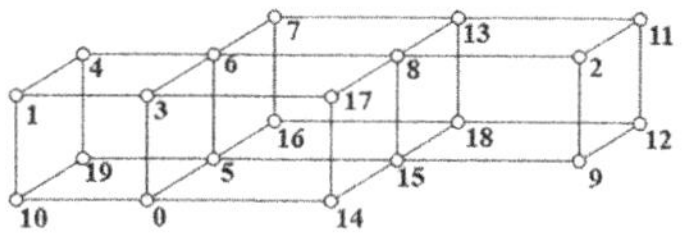


Puzzle 10

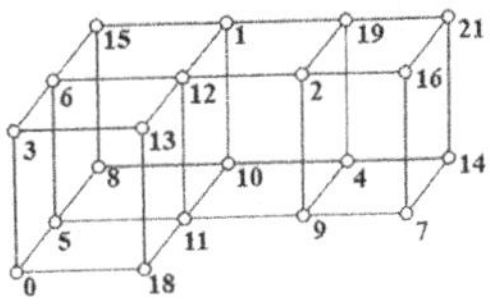


Puzzle 11

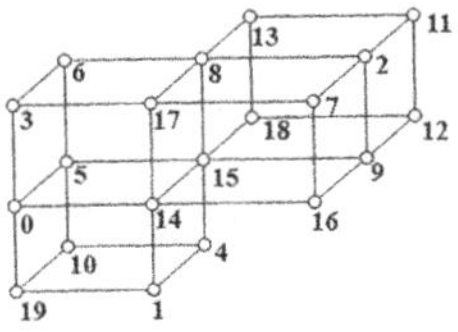


Figure 12

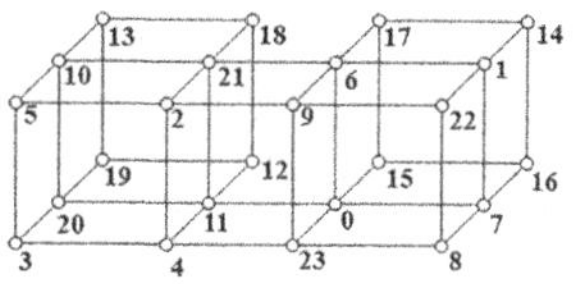


Puzzle 13

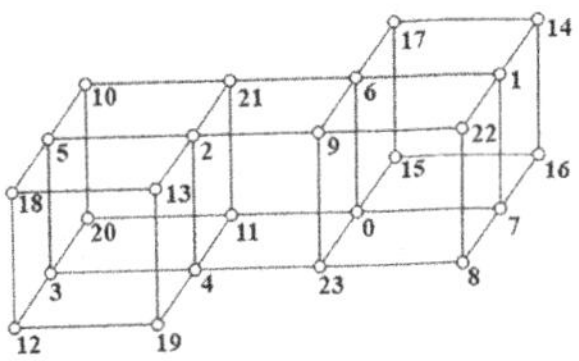


Puzzle 14

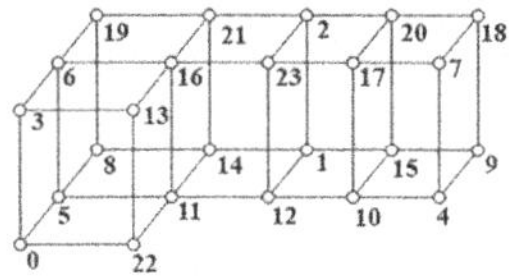


Puzzle 15

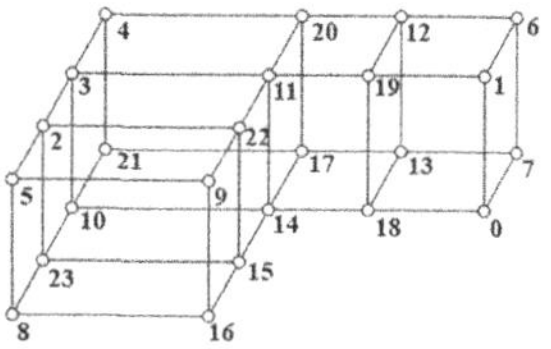

Puzzle 16

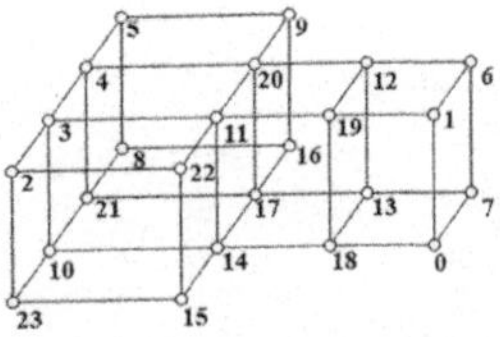


Puzzle 17

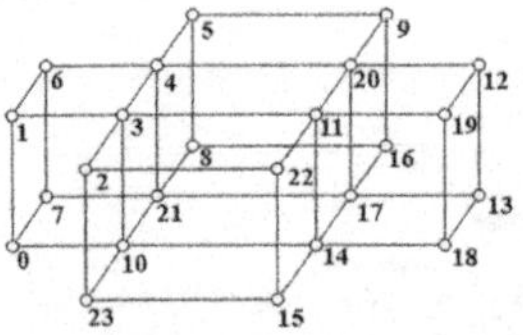


Puzzle 18

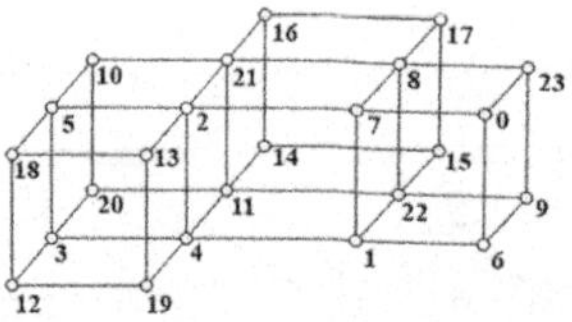


Puzzle 19

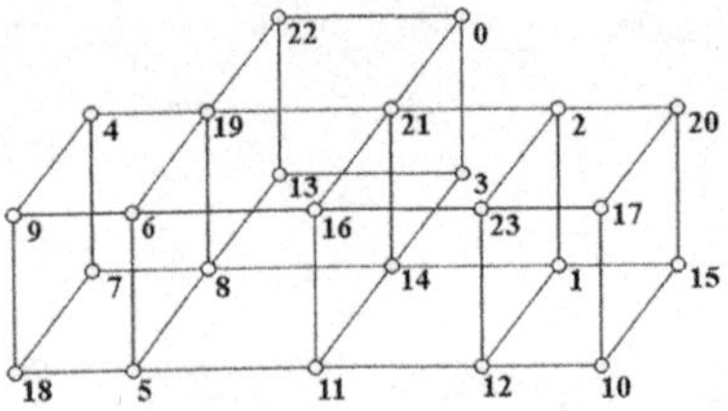


Puzzle 20

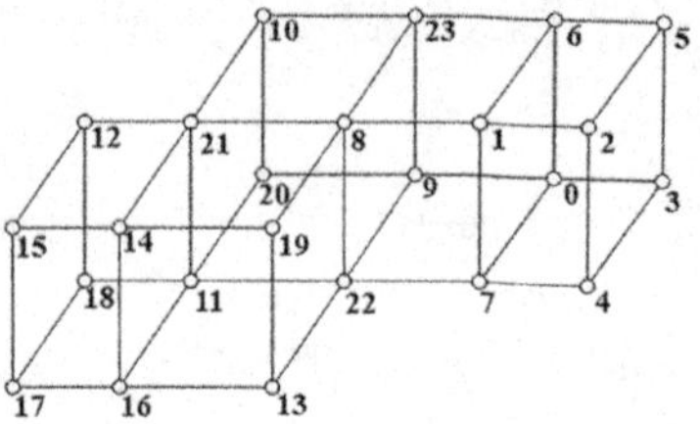


Puzzle 21

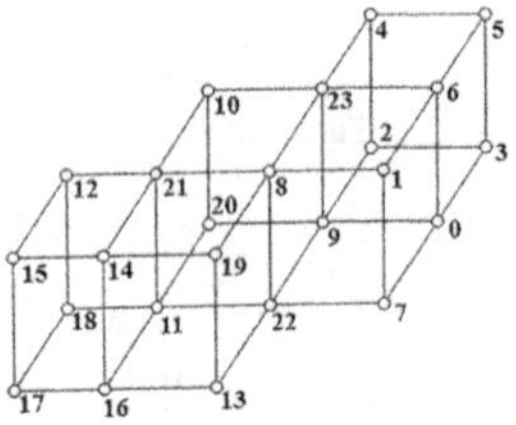


Puzzle 22

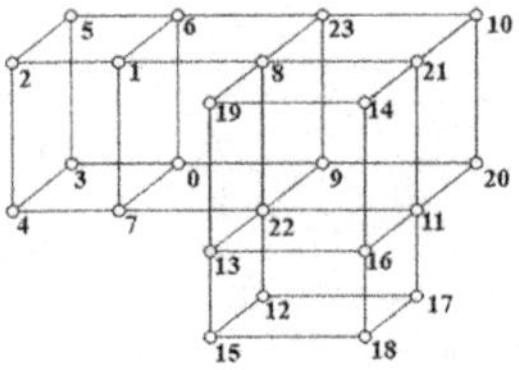


Puzzle 23

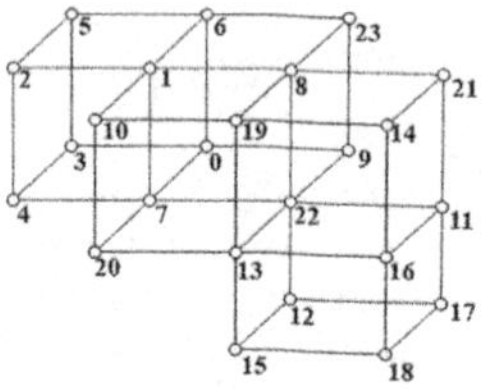


Puzzle 24

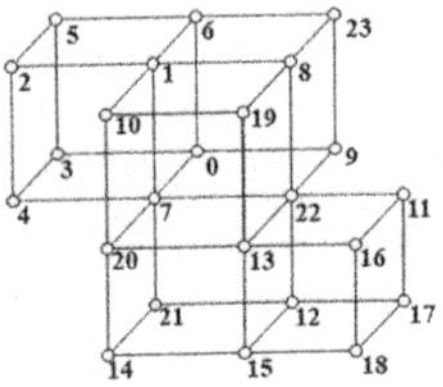


Puzzle 25

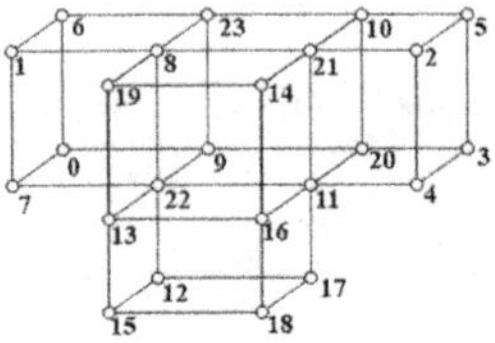


Puzzle 26

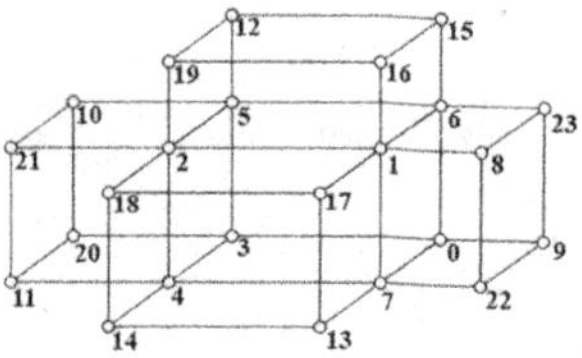

Puzzle 27

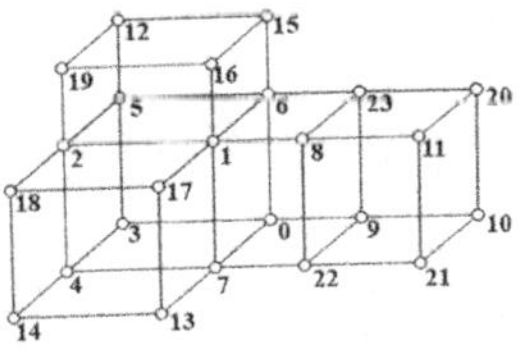


Puzzle 28

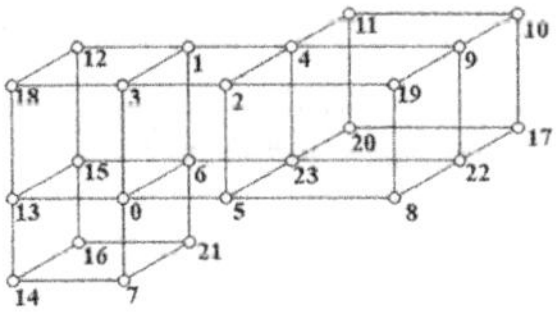


Puzzle 29

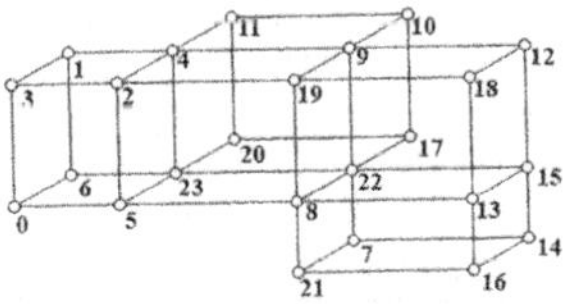


Puzzle 30

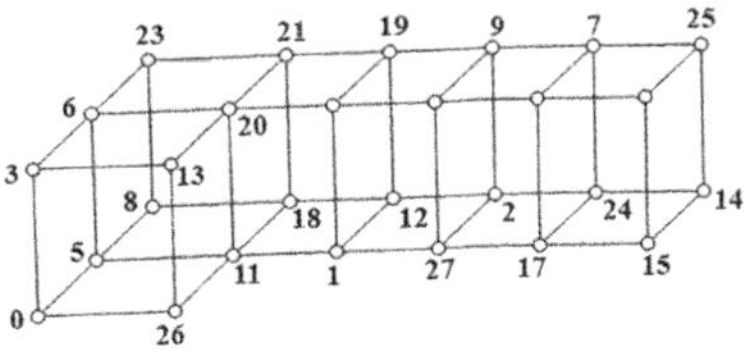


Puzzle 31

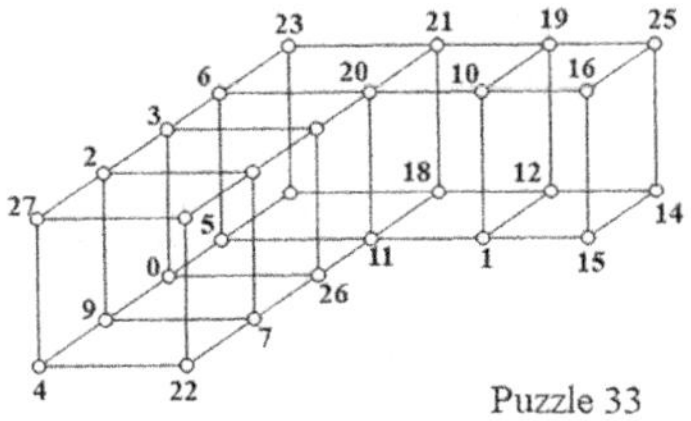


Puzzle 32

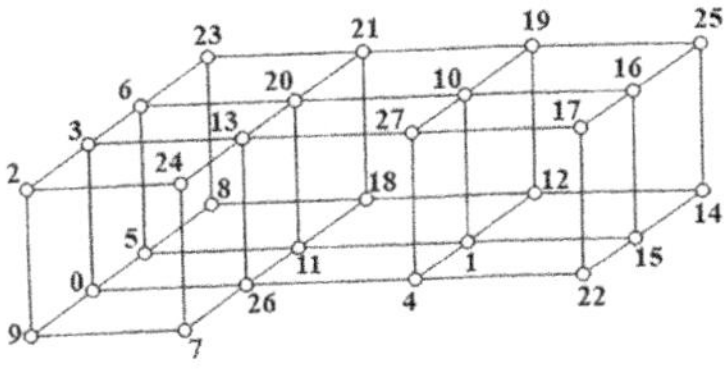


Puzzle 33

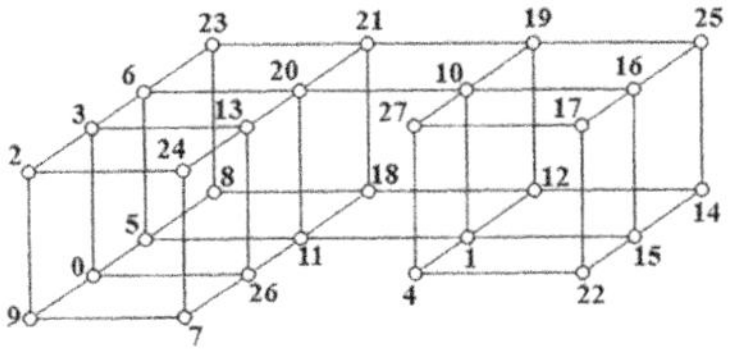


Puzzle 34

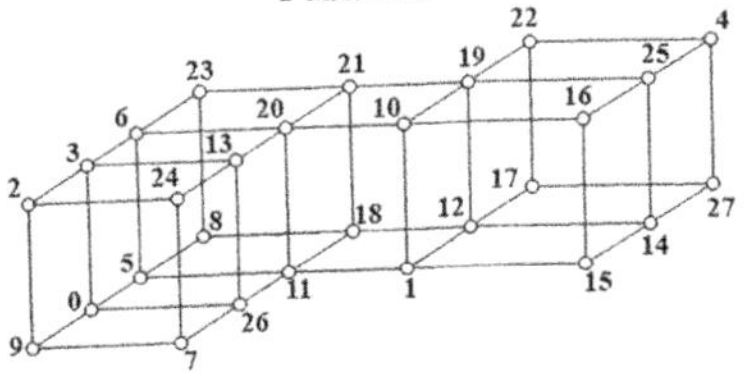


Puzzle 35

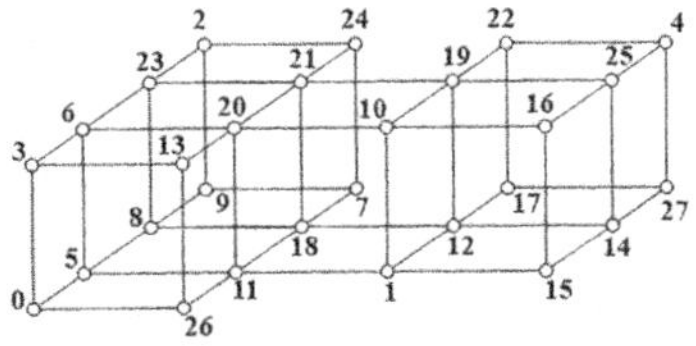


Puzzle 36

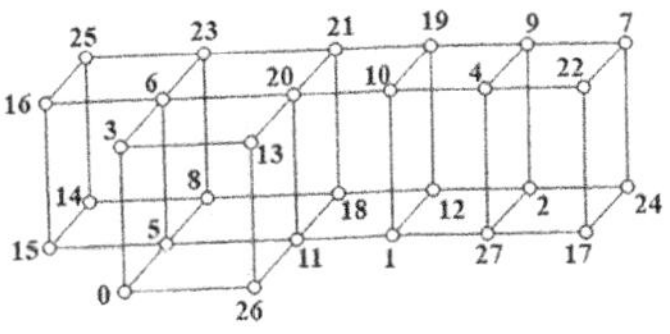


Puzzle 37

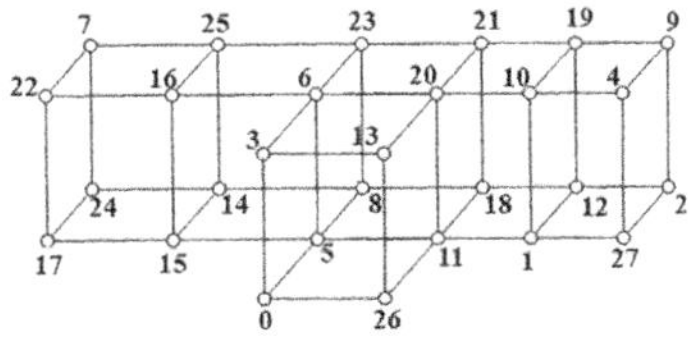

Puzzle 38

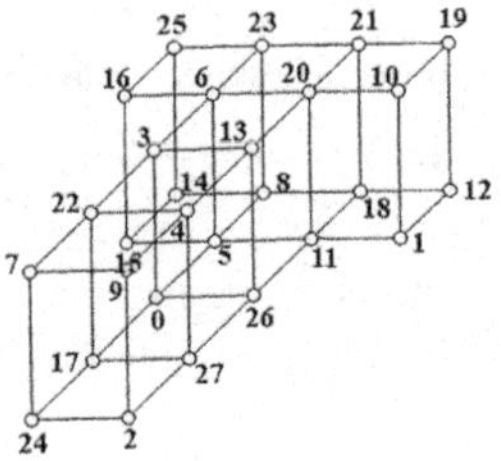

Puzzle 39

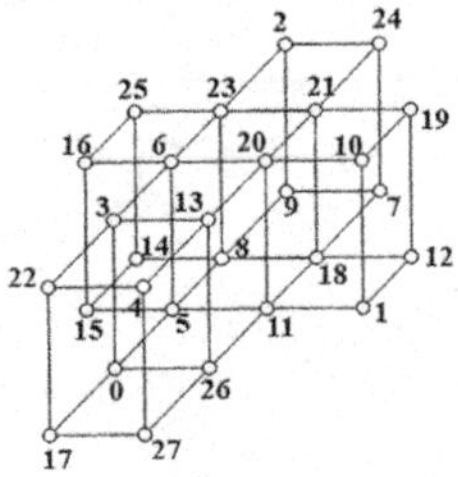

Puzzle 40

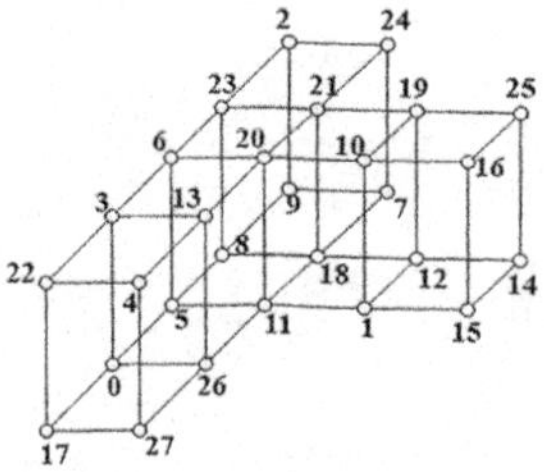

Puzzle 41

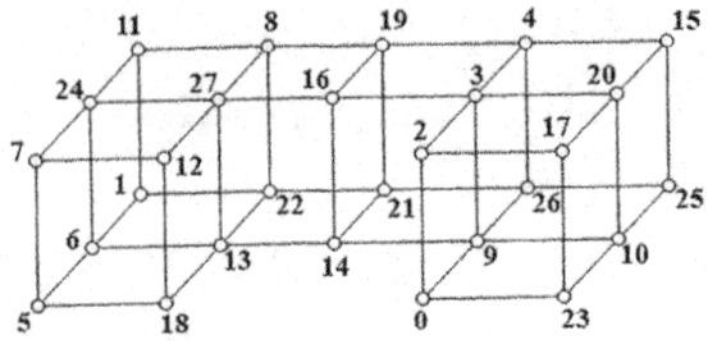

Puzzle 42

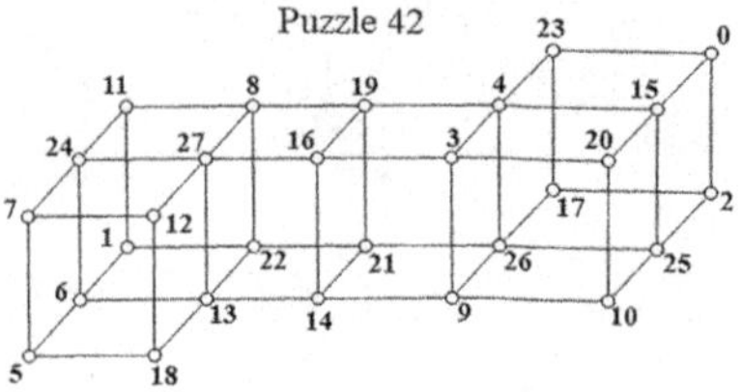

Puzzle 43

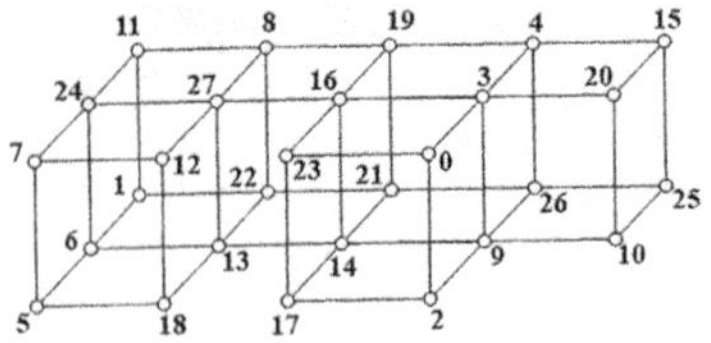

Puzzle 44

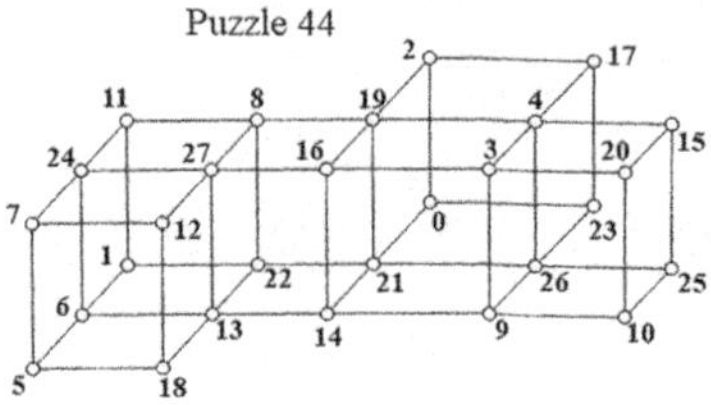

Puzzle 45

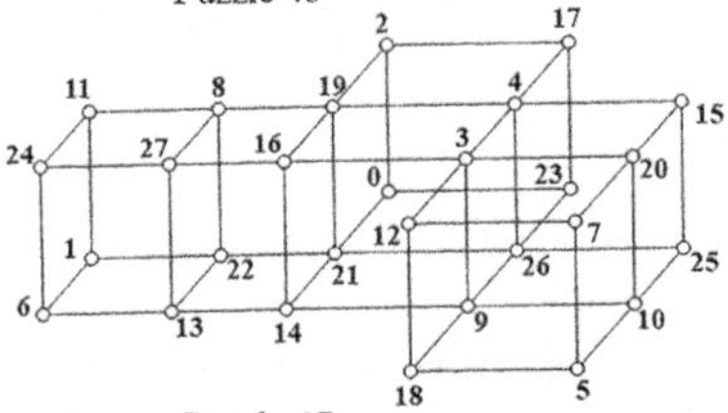

Puzzle 46

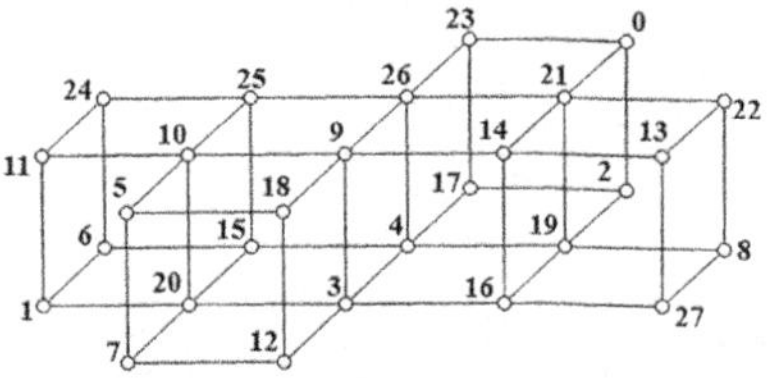

Puzzle 47

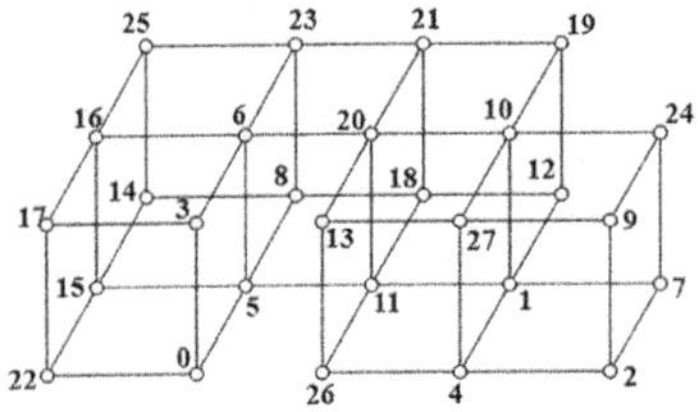

Puzzle 48

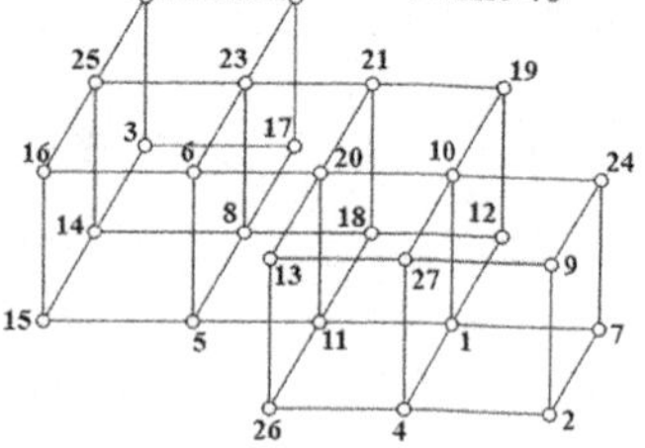

Puzzle 49

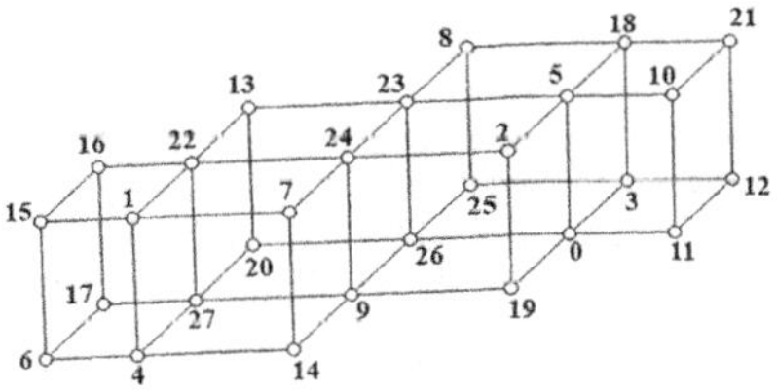


Puzzle 50

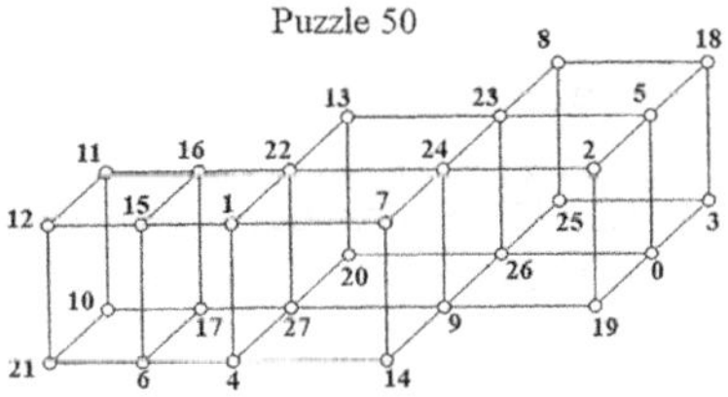


Puzzle 51

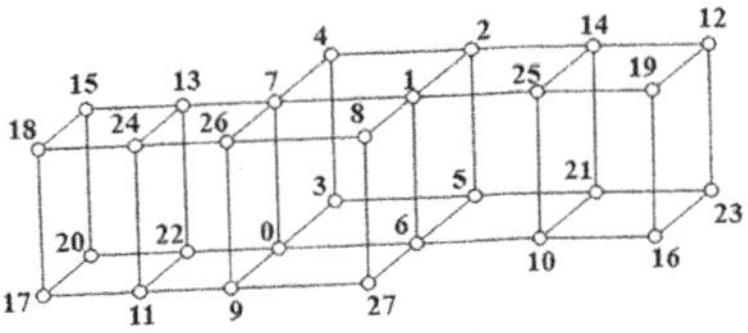


Puzzle 52

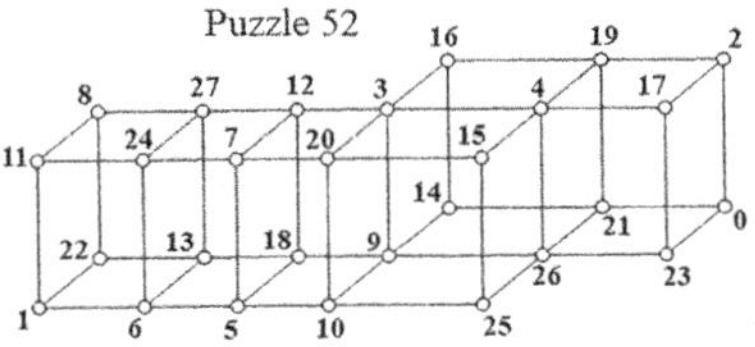


Puzzle 53

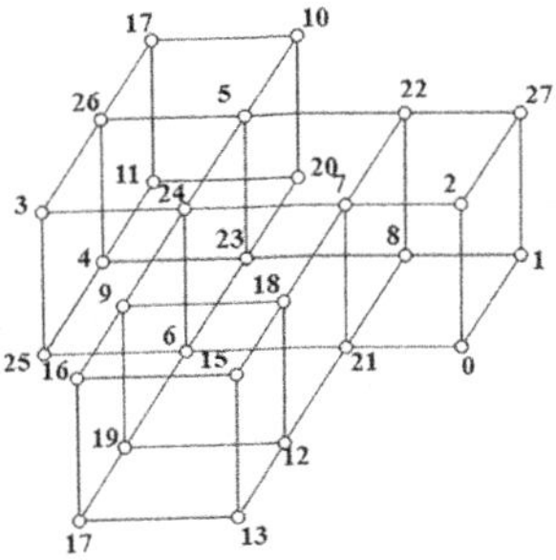


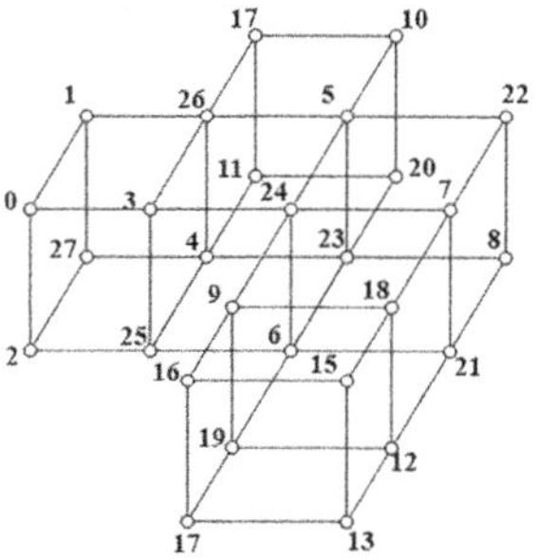


Puzzle 54

Puzzle 55

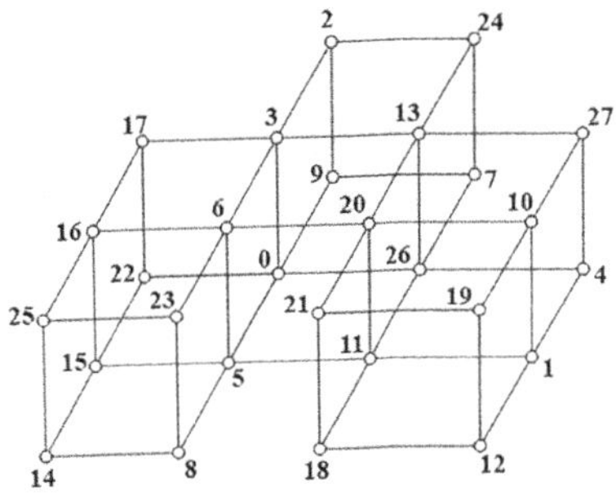


Puzzle 56

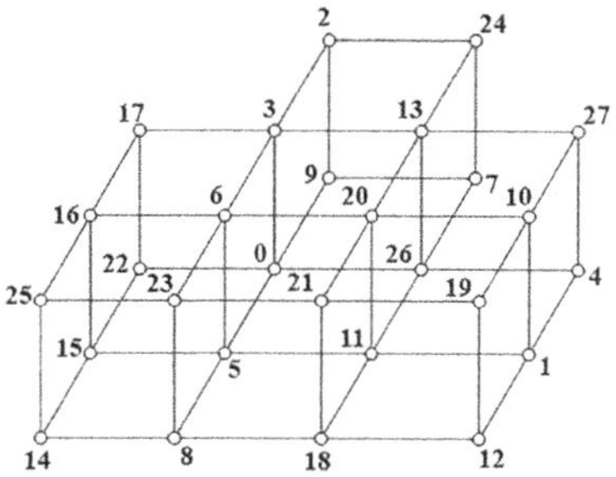

Puzzle 57

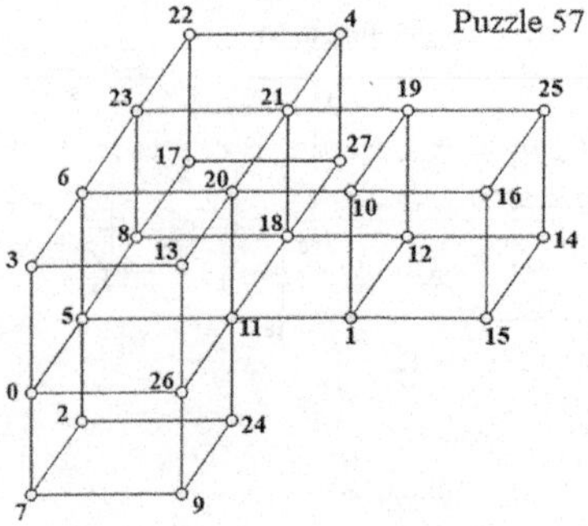


Puzzle 58

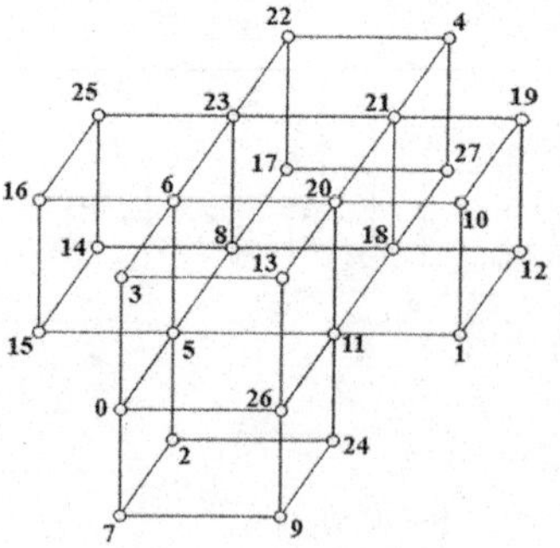


Puzzle 59

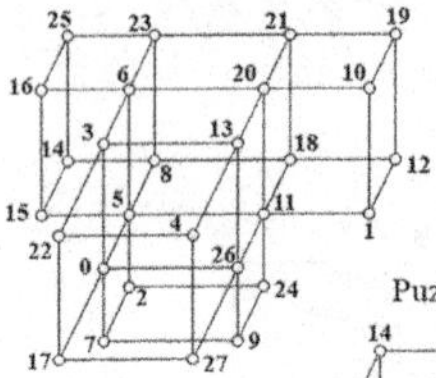


Puzzle 60

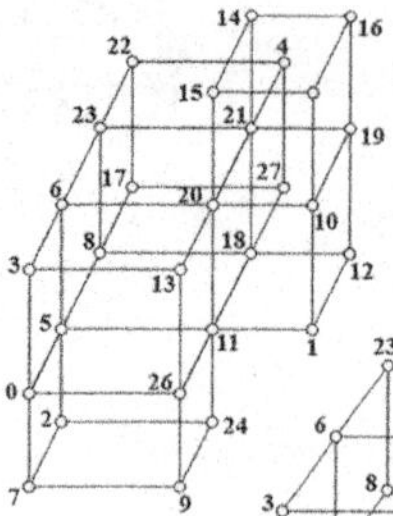


Puzzle 61

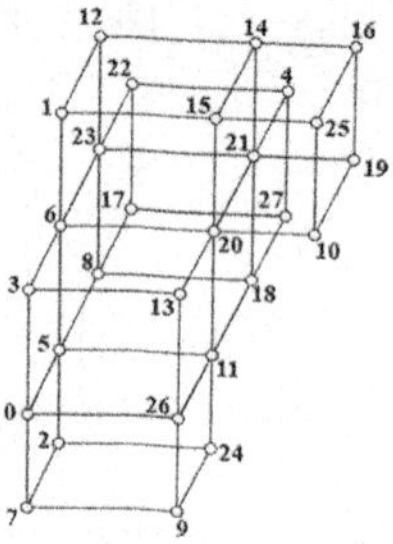


Puzzle 62

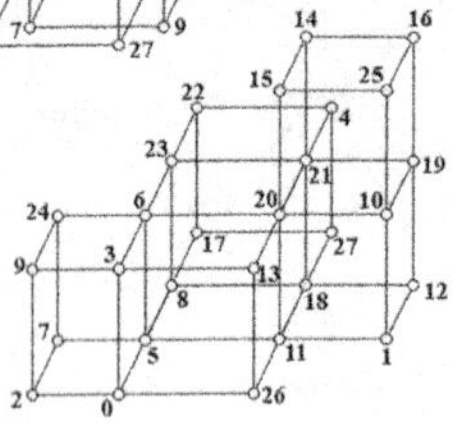


Puzzle 63

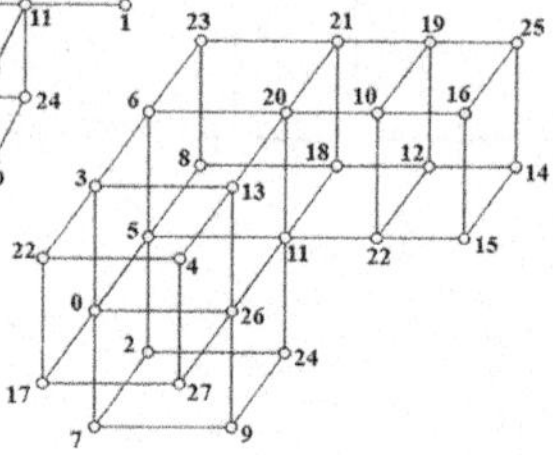


Puzzle 64

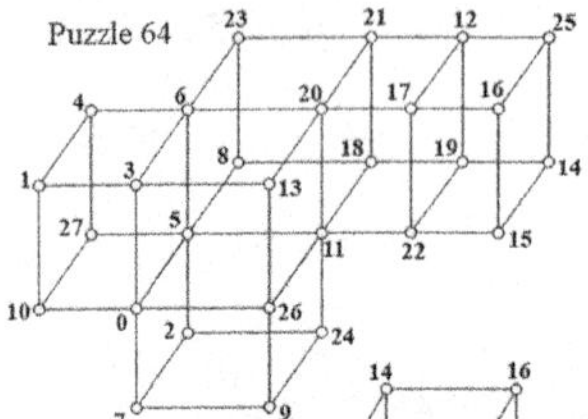


Puzzle 65

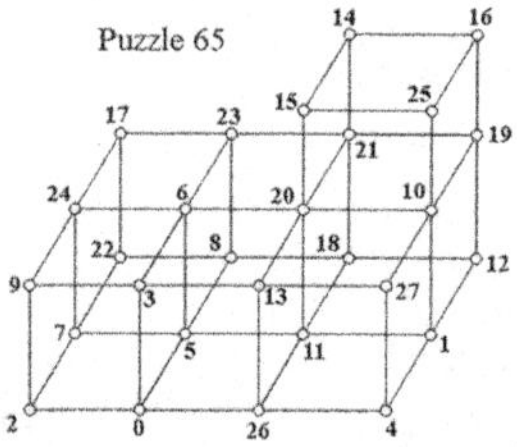


Puzzle 66

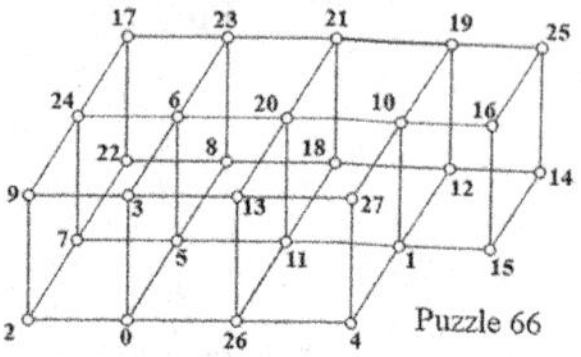


Puzzle 67

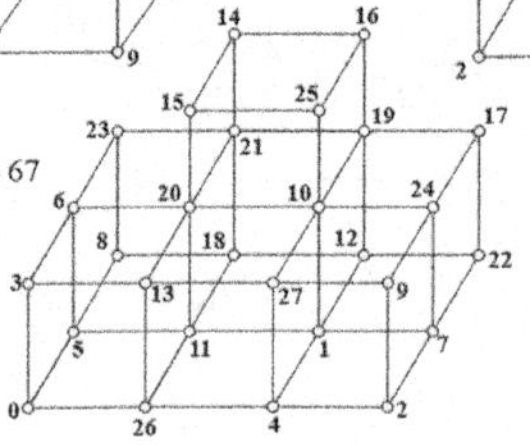


Puzzle 68

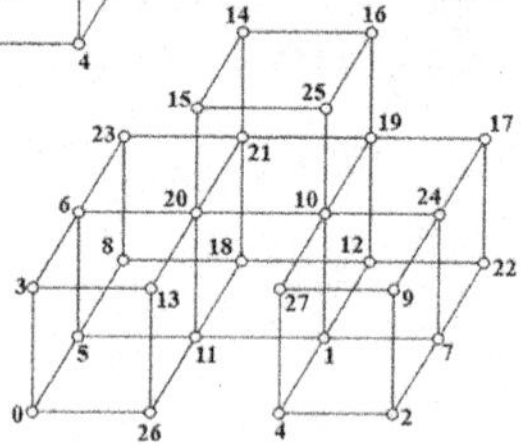

Puzzle 69

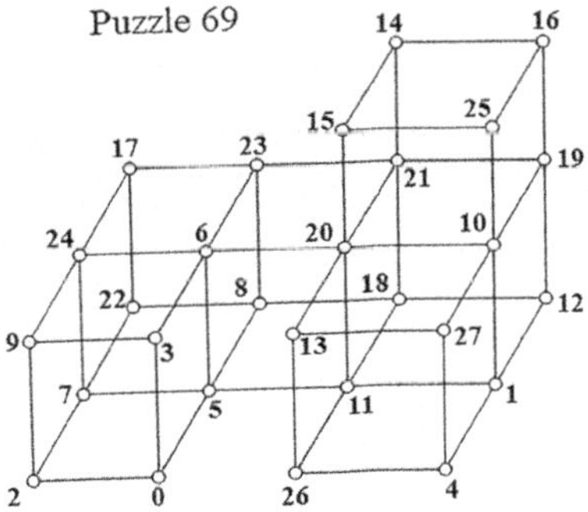


Puzzle 70

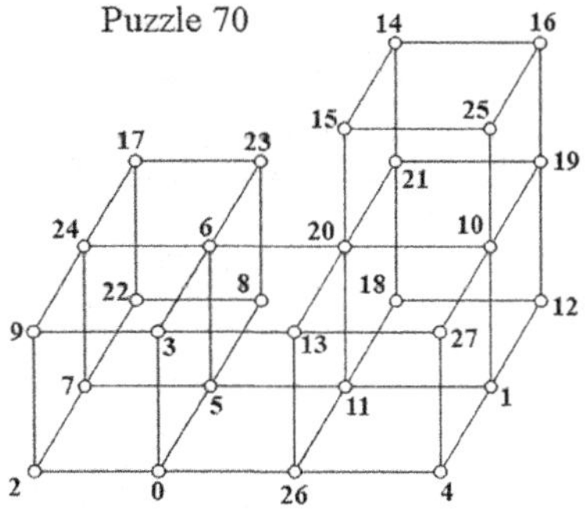


Puzzle 71

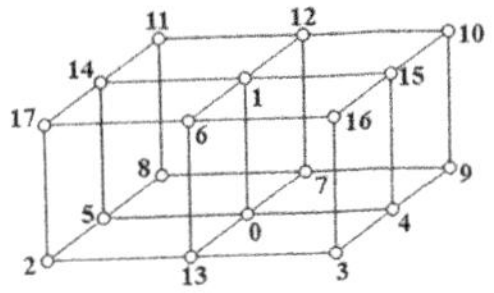


Puzzle 72

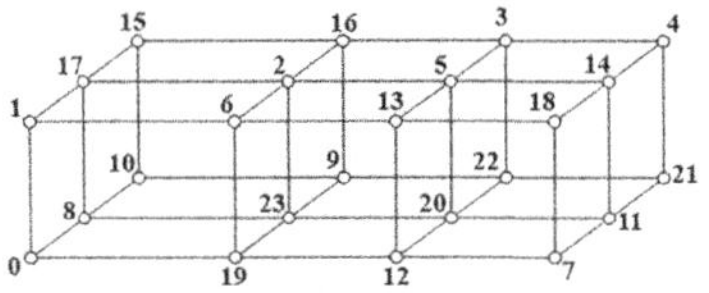


Puzzle 73

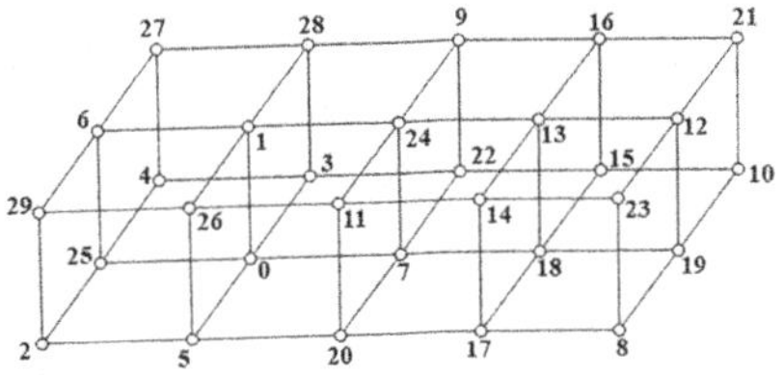


Puzzle 74

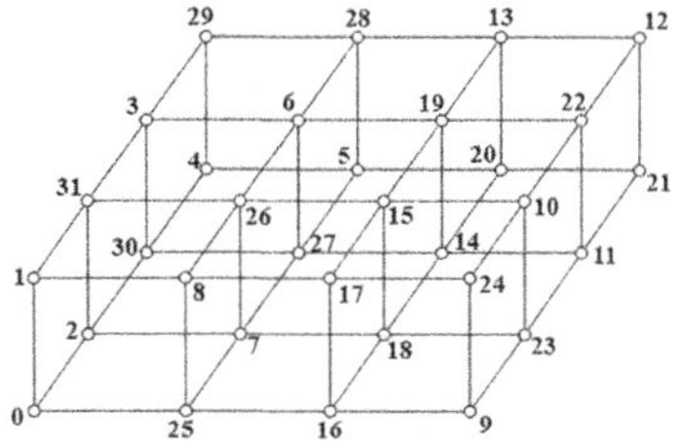


Puzzle 75

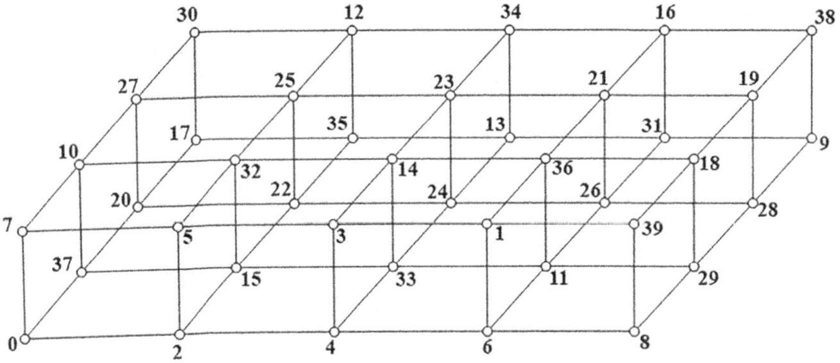

Puzzle 76

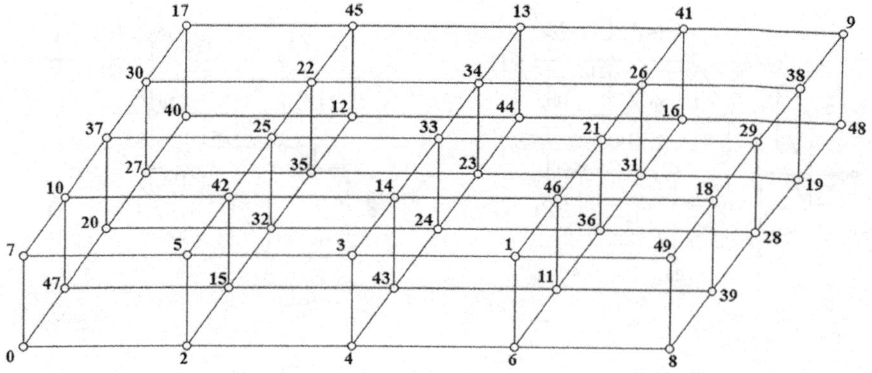


Puzzle 77

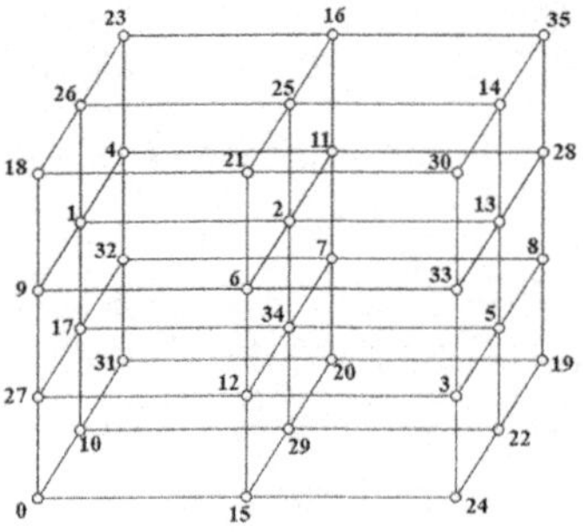


Puzzle 78

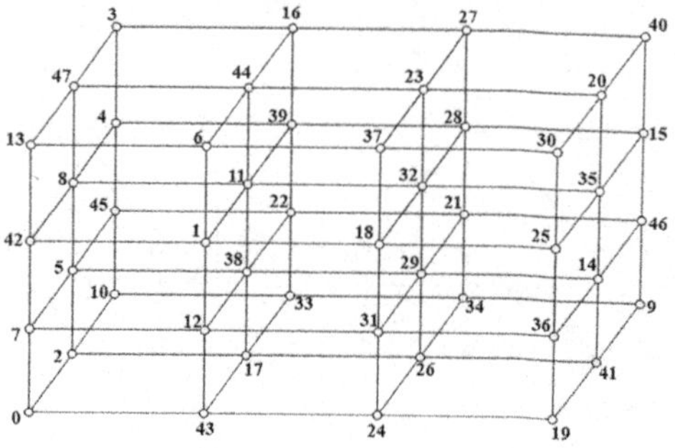

## Puzzle 79

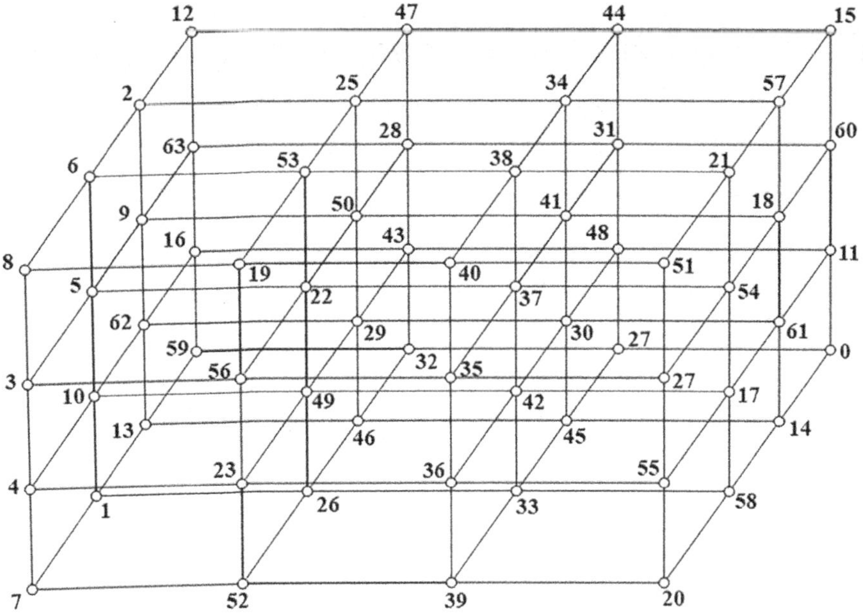


Puzzle 80

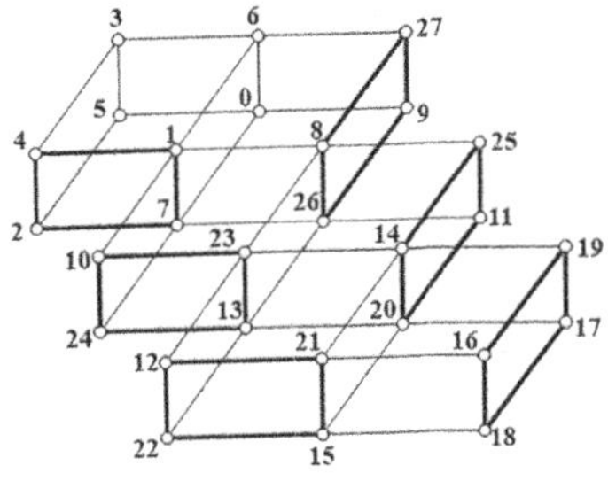


Puzzle 81

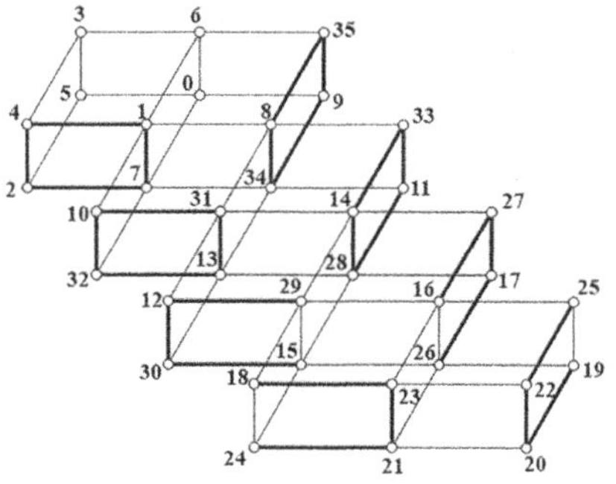

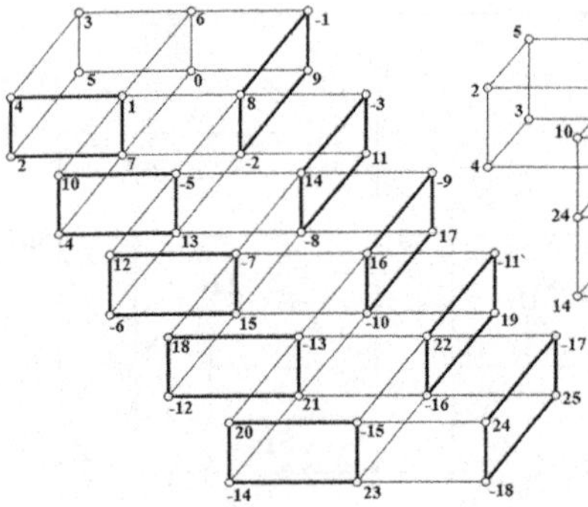

Puzzle 82

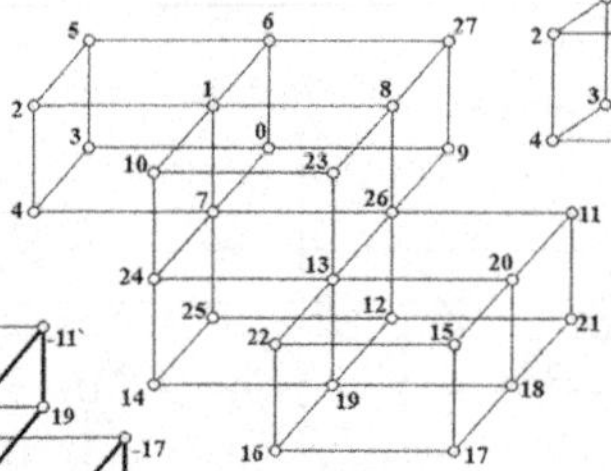

Puzzle 83

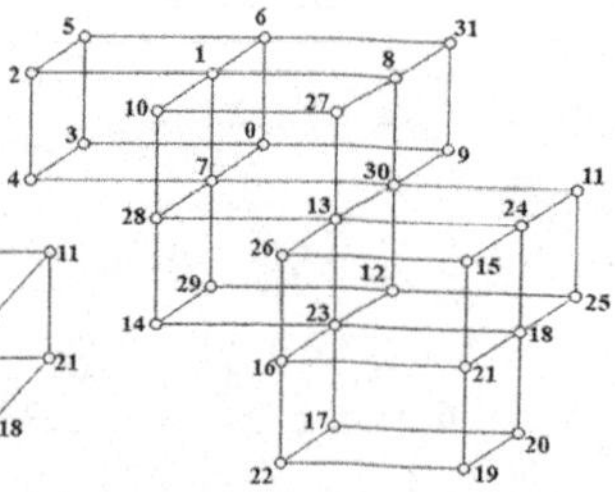

Puzzle 84

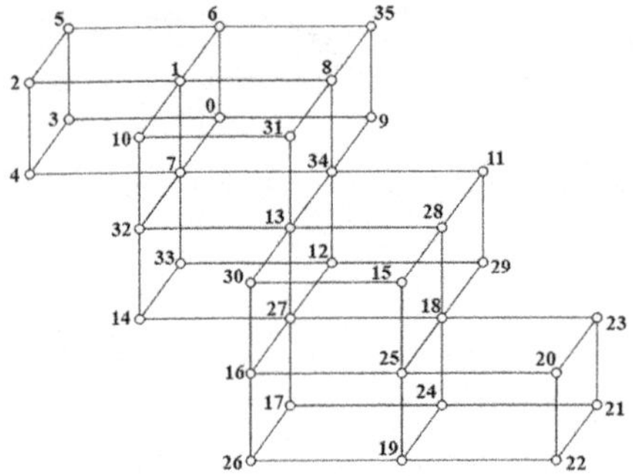

Puzzle 85

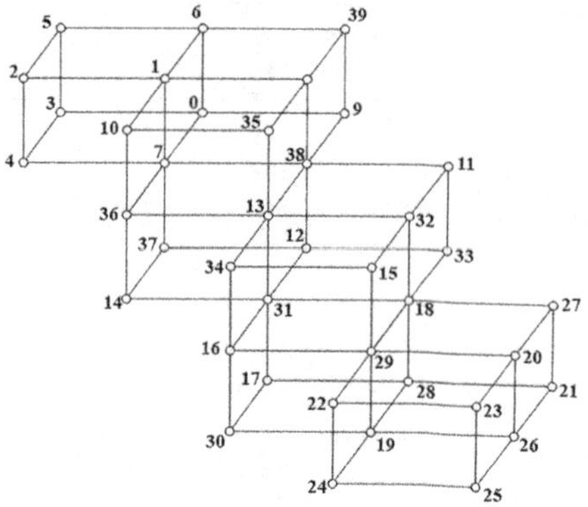

Puzzle 86

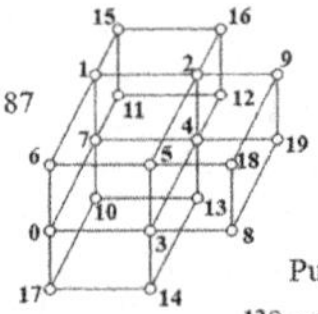

Puzzle 87

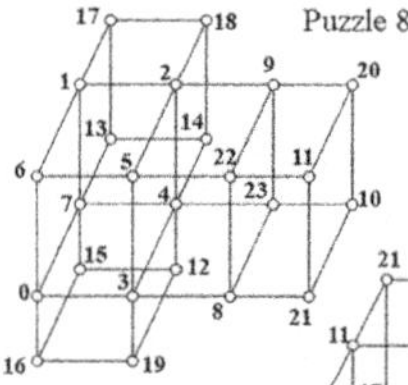

Puzzle 88

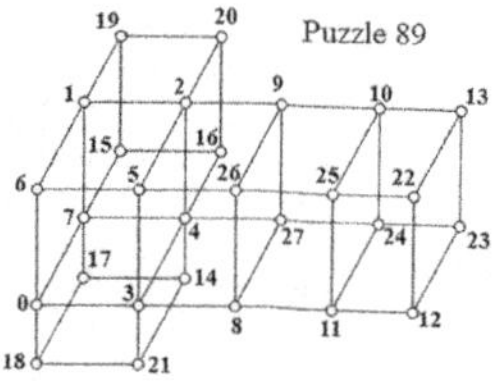

Puzzle 89

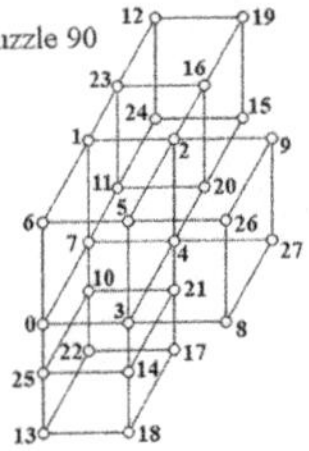

Puzzle 90

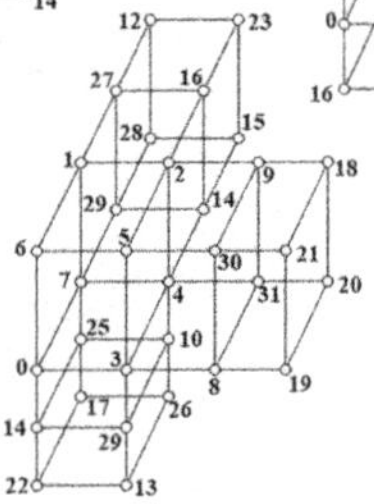

Puzzle 91

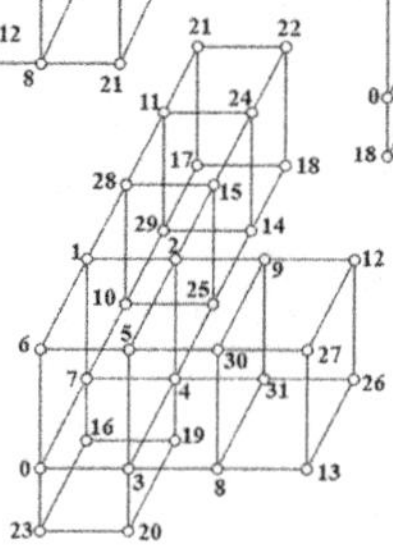

Puzzle 92

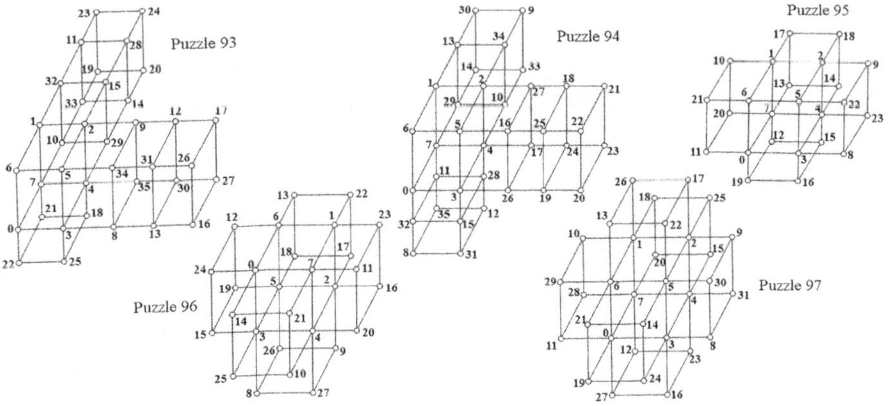


Puzzle 98

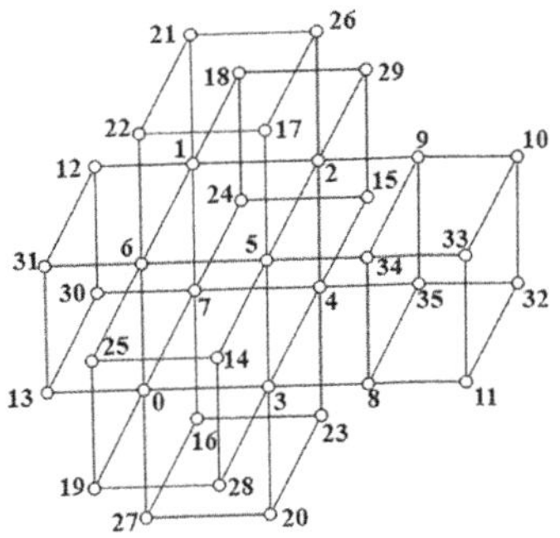


Puzzle 99

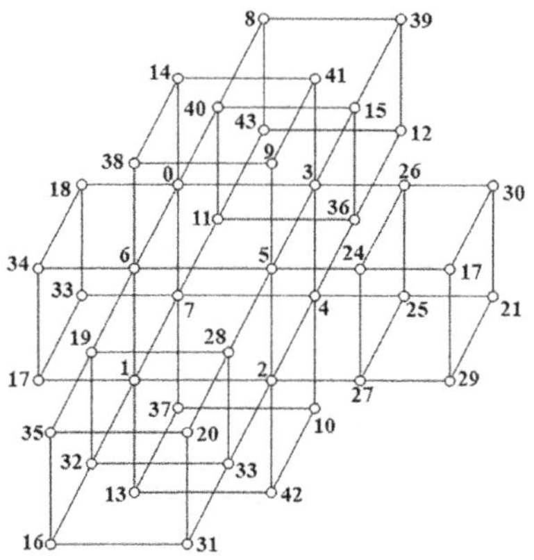

Puzzle 100

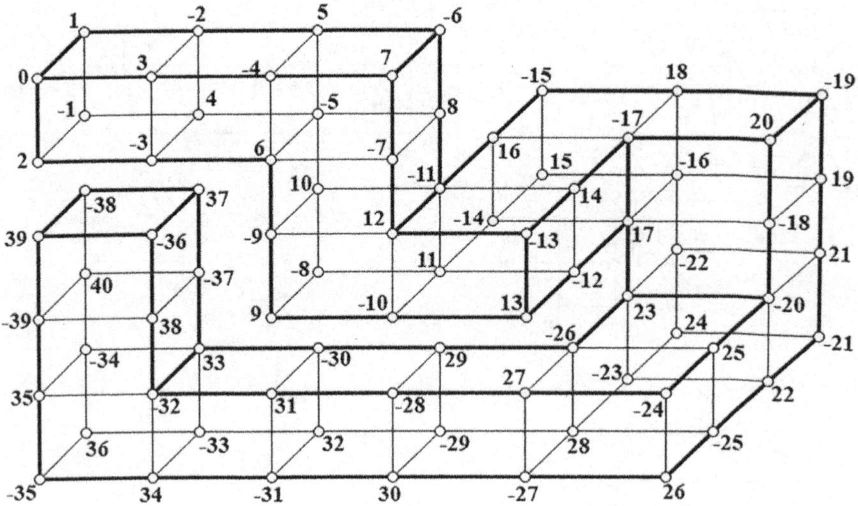
1
-2
5
-6
0
3
-4
7
-15
18
-19
-1
4
-5
8
-17
20
2
-3
6
-7
16
15
-16
19
10
-11
14
-38
37
39
-36
12
-14
-13
17
-18
-9
21
-37
-8
11
-12
-22
40
9
-10
13
23
-20
-39
38
24
-26
25
-21
-34
33
-30
29
-23
27
22
35
-32
31
-28
-24
36
-33
32
-29
28
-25
-35
34
-31
30
-27
26

# Triangulation Puzzle Solutions

Puzzle 101

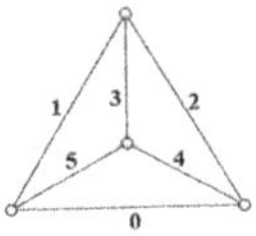


Puzzle 102

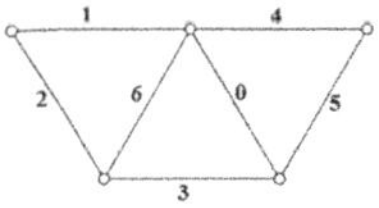


Puzzle 103

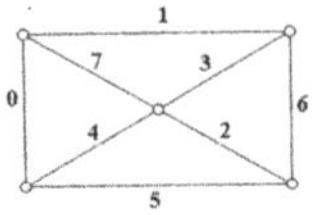


Puzzle 104

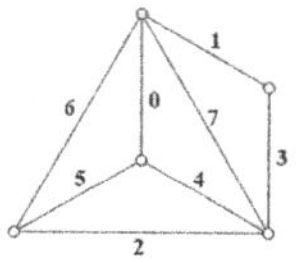


Puzzle 105

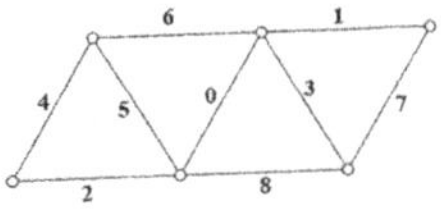


Puzzle 106

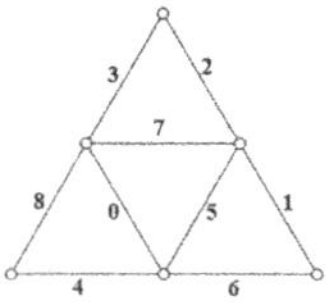


Puzzle 107

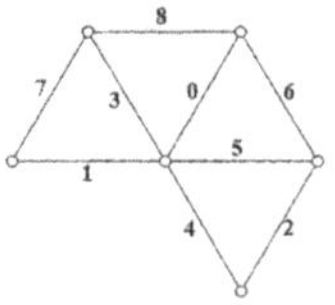


Puzzle 108

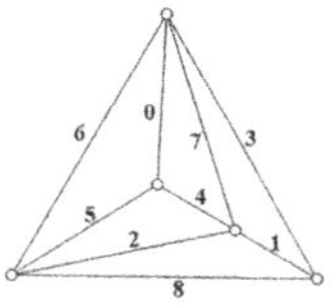


Puzzle 109

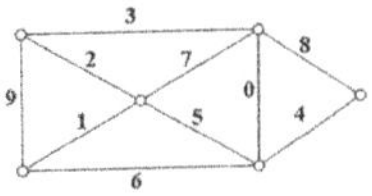


Puzzle 110

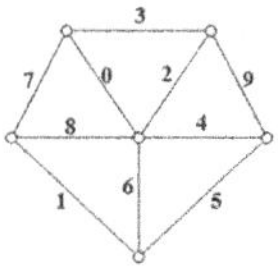


Puzzle 111

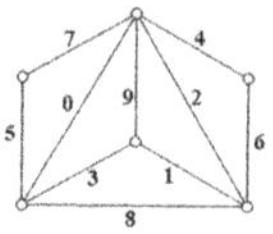


Puzzle 112

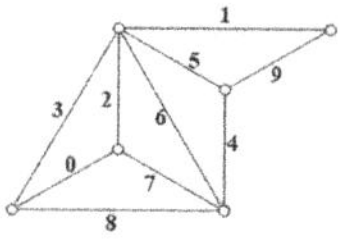


Puzzle 113

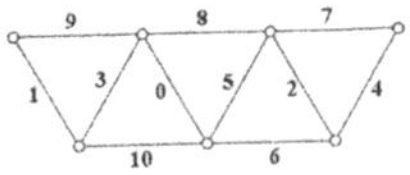


Puzzle 114

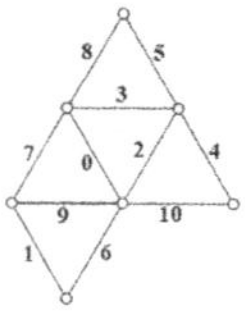


Puzzle 115

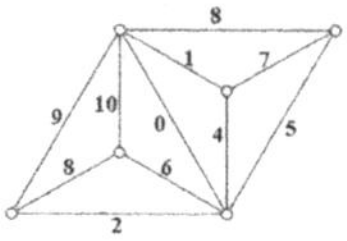


Puzzle 116

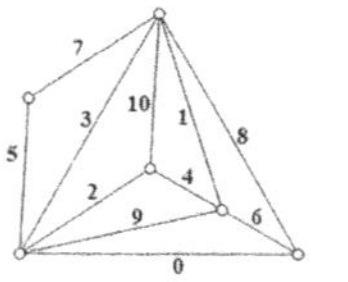


Puzzle 117

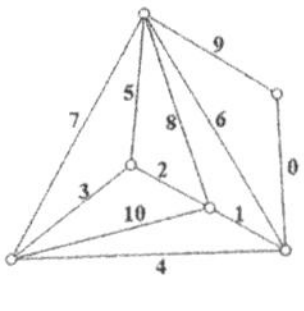


Puzzle 118

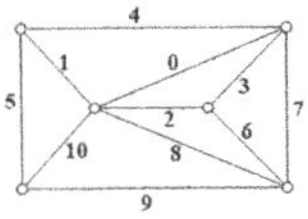


Puzzle 119

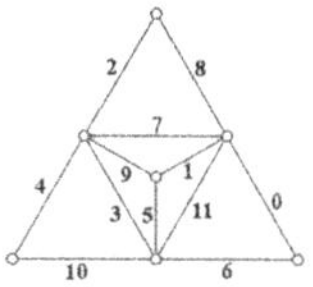


Puzzle 120

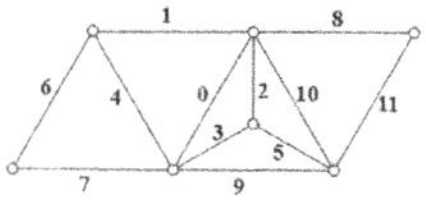


Puzzle 121

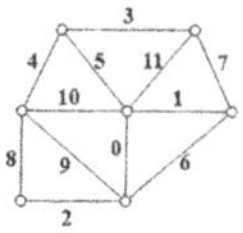


Puzzle 122

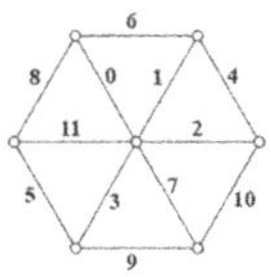


Puzzle 123

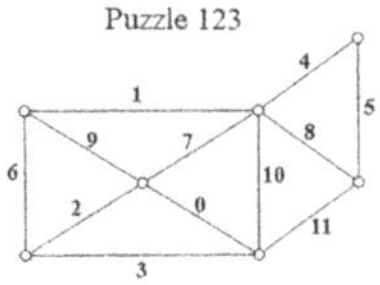


Puzzle 124

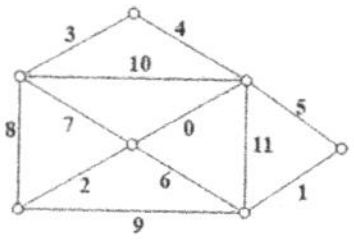

Puzzle 125

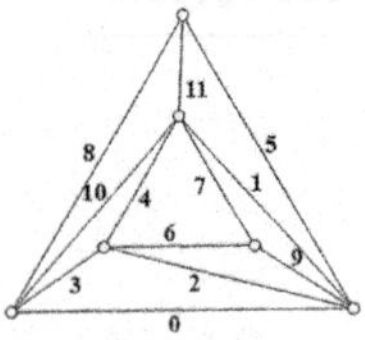


Puzzle 126

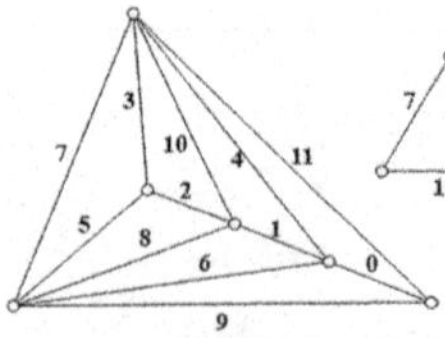


Puzzle 127

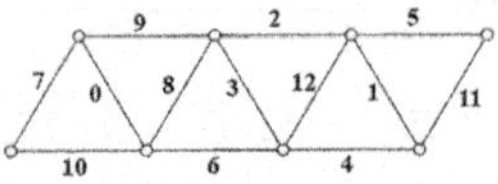


Puzzle 128

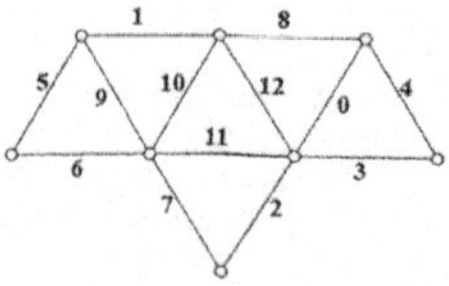


Puzzle 129

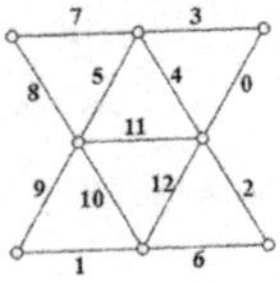


Puzzle 130

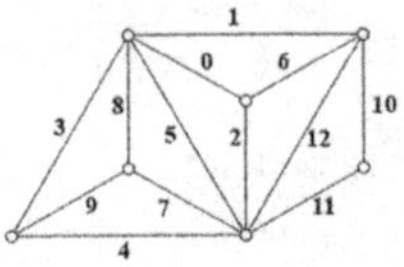


Puzzle 131

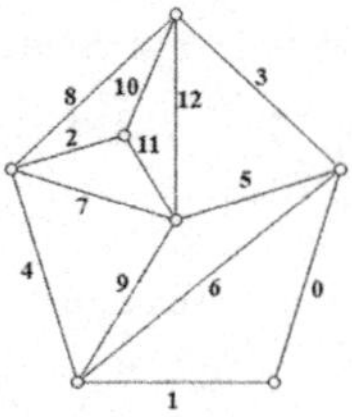


Puzzle 132

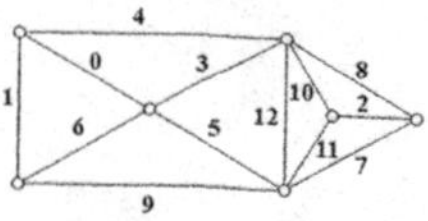


Puzzle 133

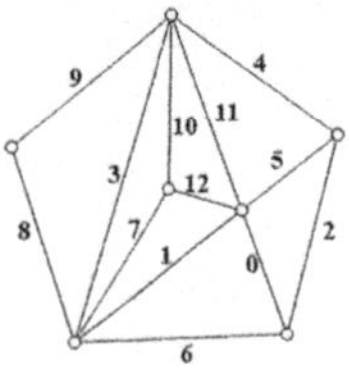


Puzzle 134

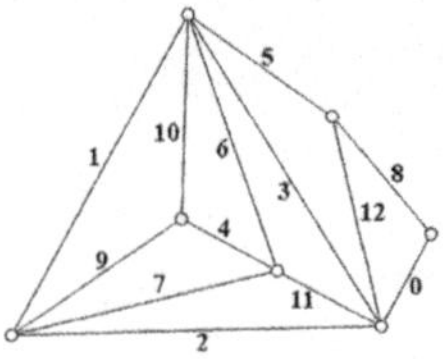


Puzzle 135

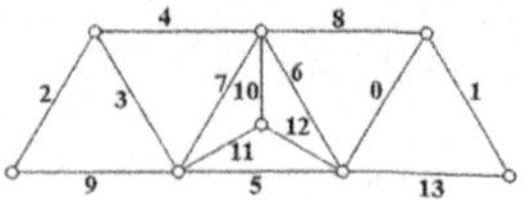


Puzzle 136

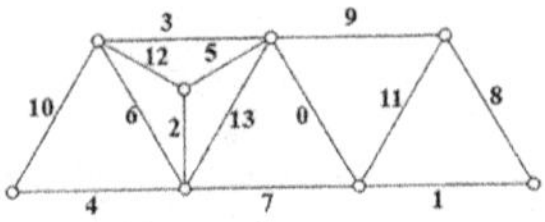


Puzzle 137

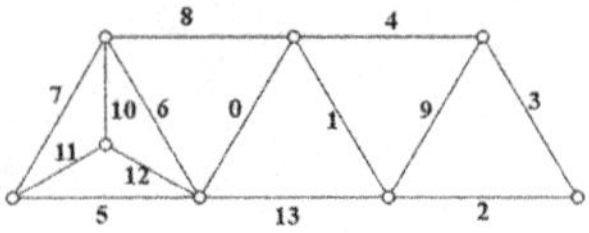


Puzzle138

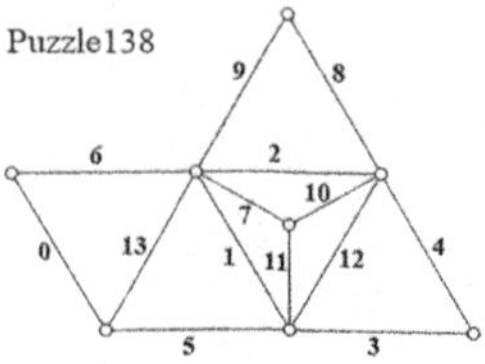


Puzzle 139

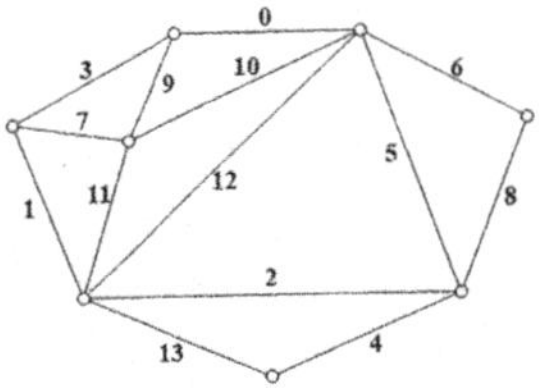


Puzzle 140

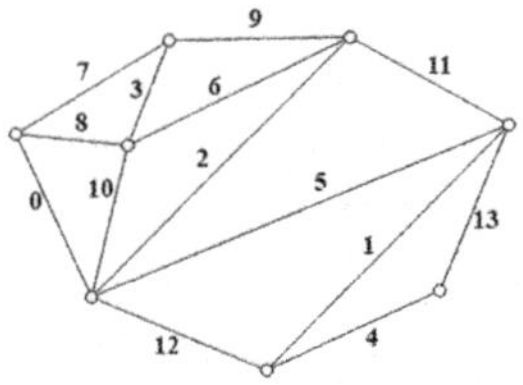


Puzzle 141

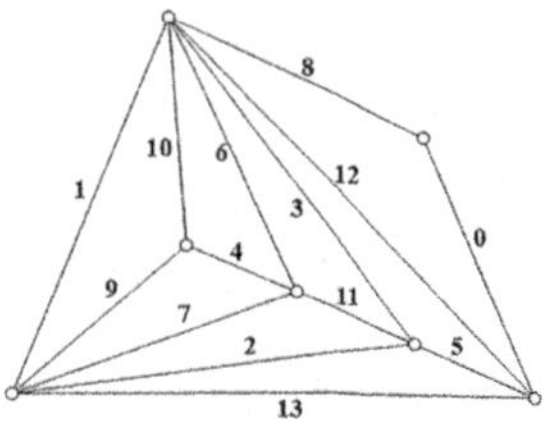

Puzzle 142

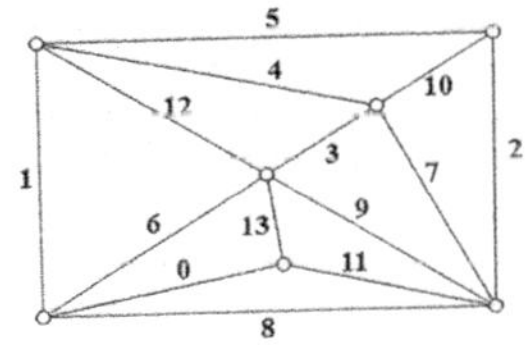

Puzzle 143

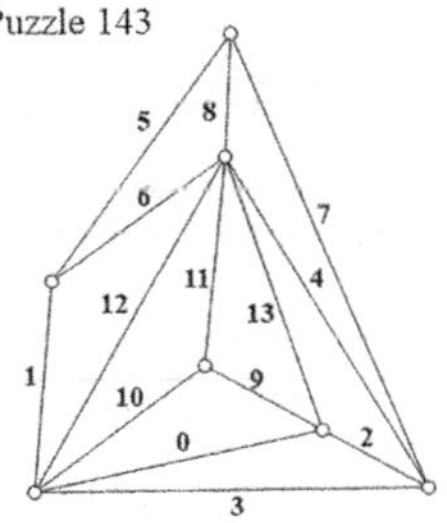

Puzzle 144

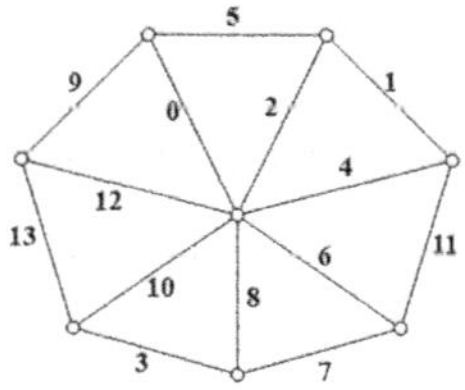

Puzzle 145

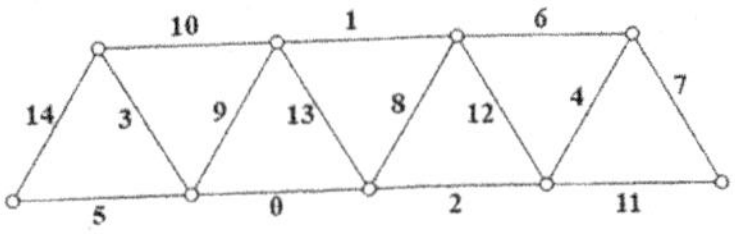

Puzzle 146

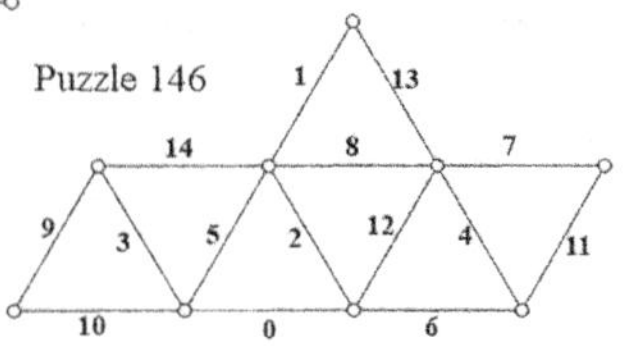

Puzzle 147

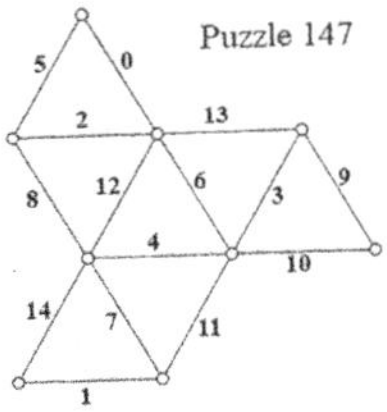

Puzzle 148

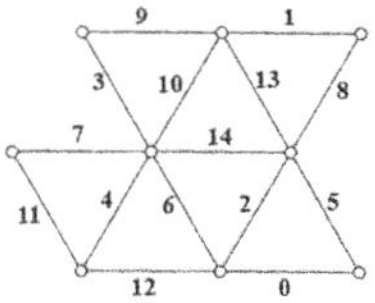

Puzzle 149

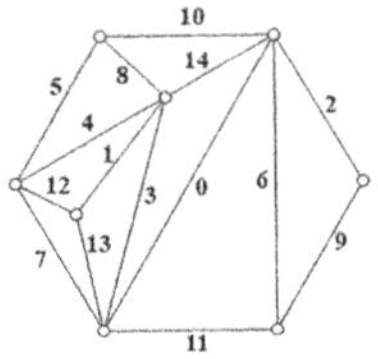

Puzzle150

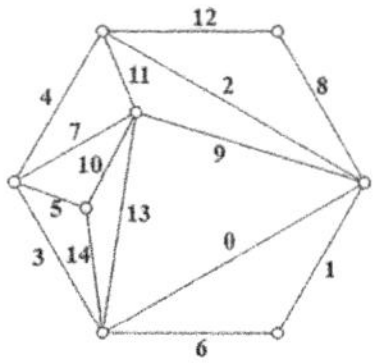

Puzzle 151

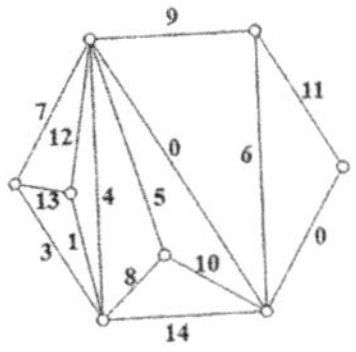

Puzzle 152

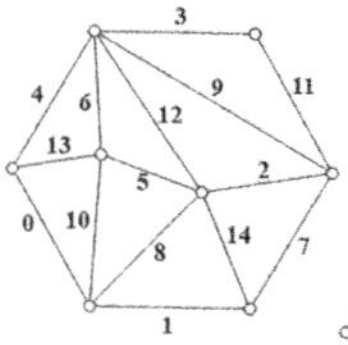

Puzzle 153

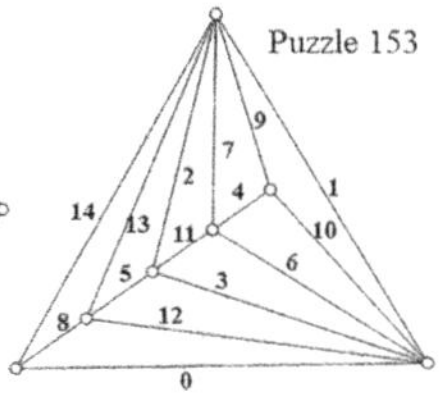

Puzzle 154

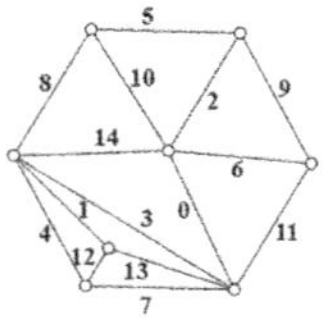

Puzzle 155

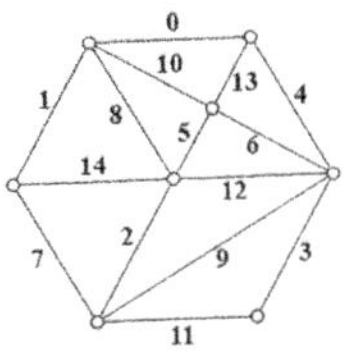

Puzzle 156

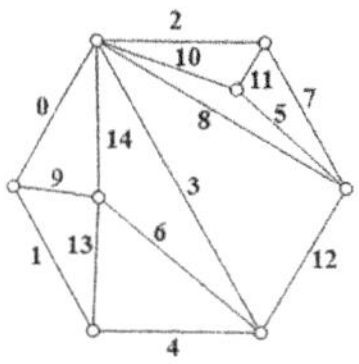

Puzzle 157

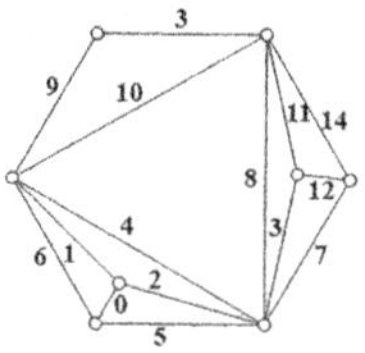

Puzzle 158

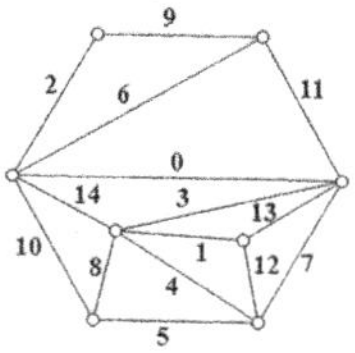

Puzzle 159

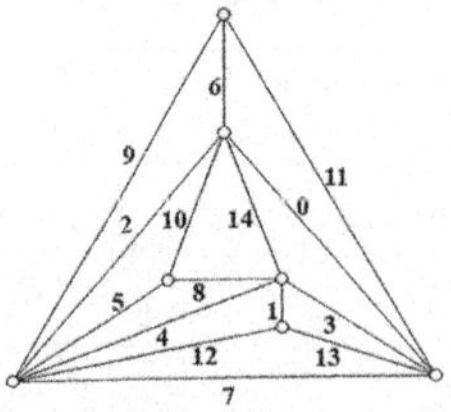


Puzzle 160

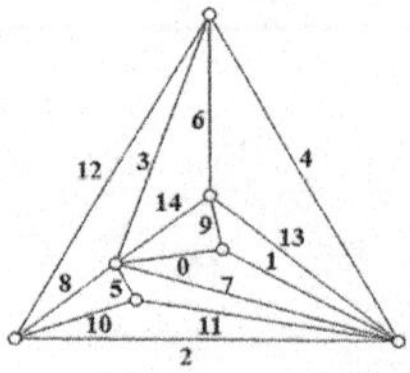


Puzzle 161

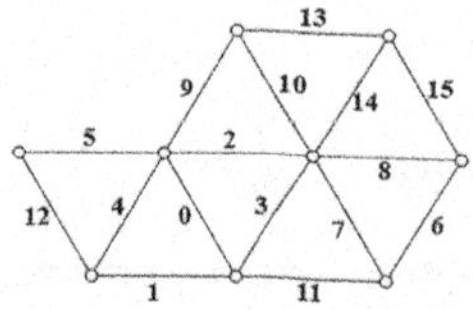


Puzzle 162

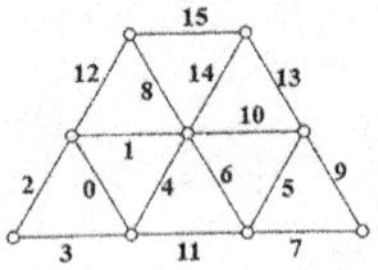


Puzzle 163

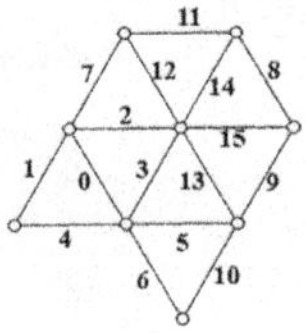


Puzzle 164

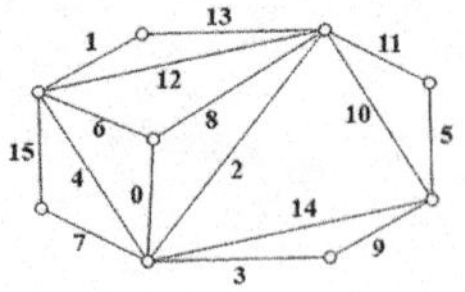


Puzzle 165

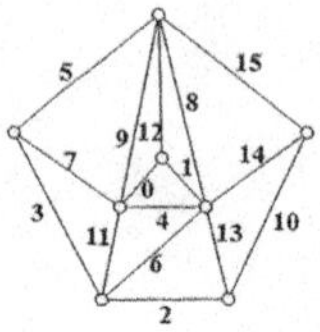


Puzzle 166

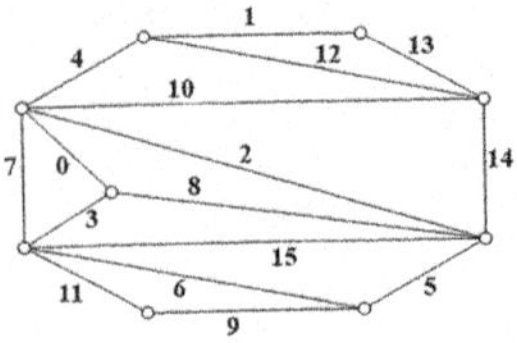


Puzzle 167

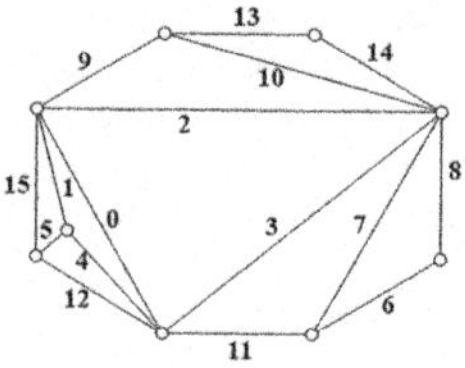


Puzzle 168

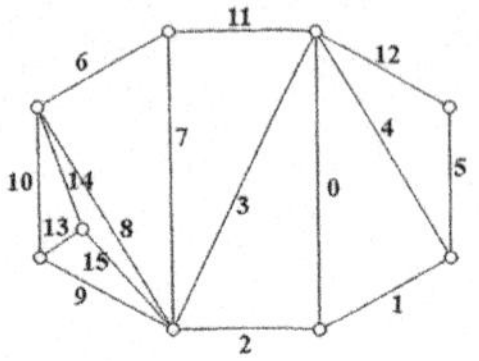


Puzzle 169

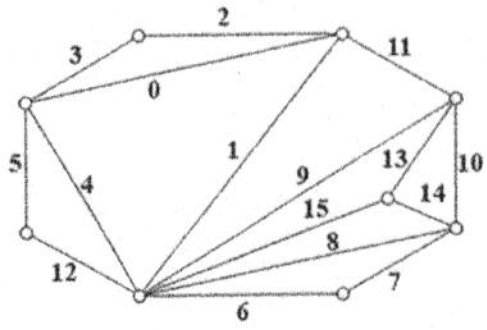


Puzzle 170

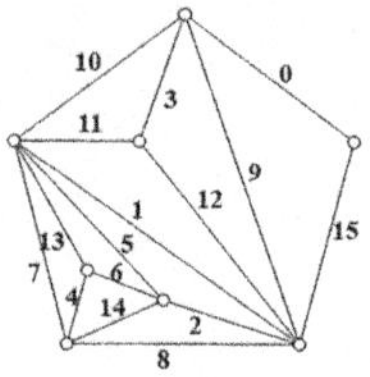


Puzzle 171

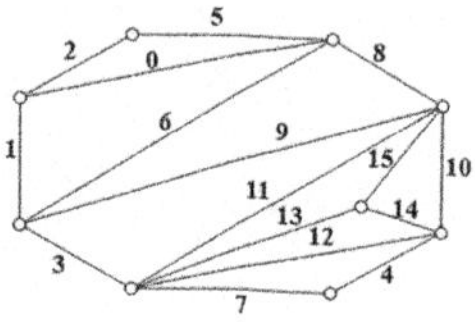


Puzzle 172

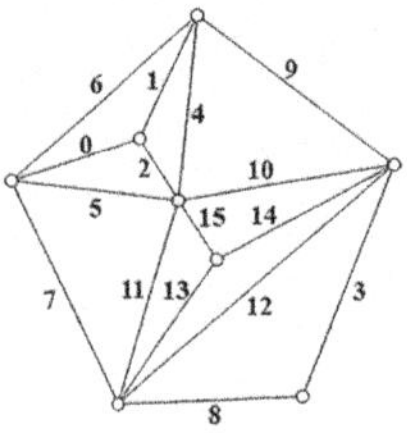


Puzzle 173

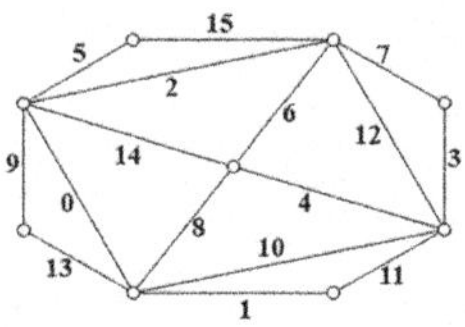


Puzzle 174

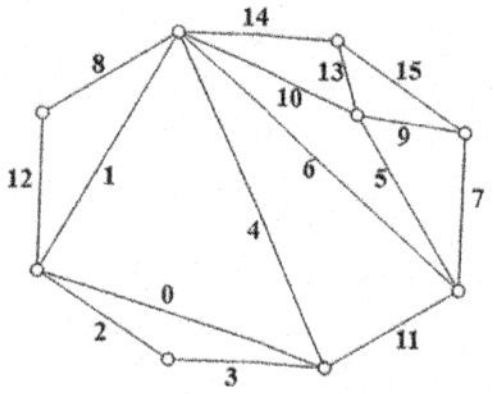

Puzzle 175

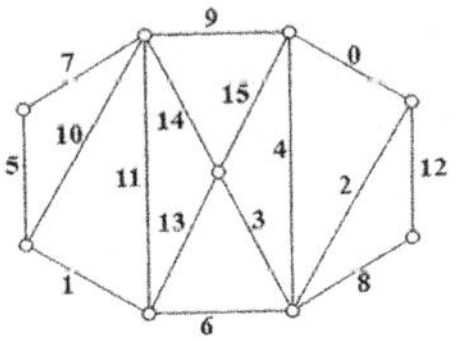


Puzzle 176

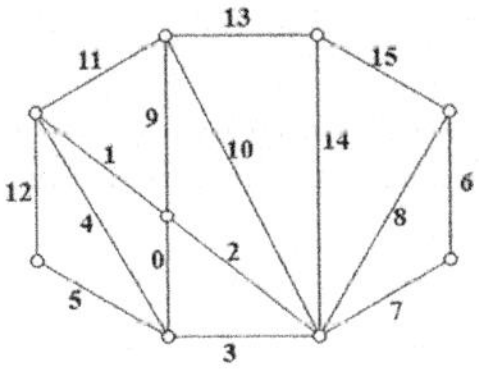


Puzzle 177

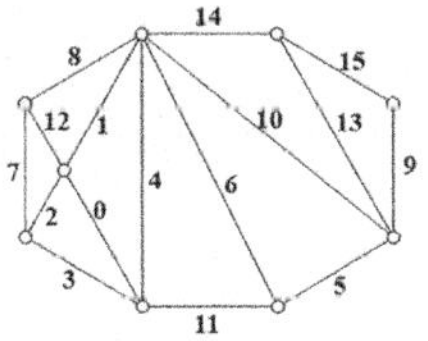


Puzzle 178

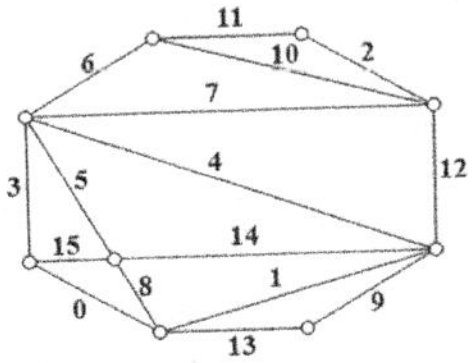


Puzzle 179

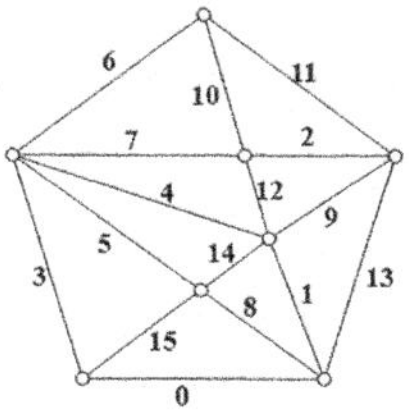


Puzzle 180

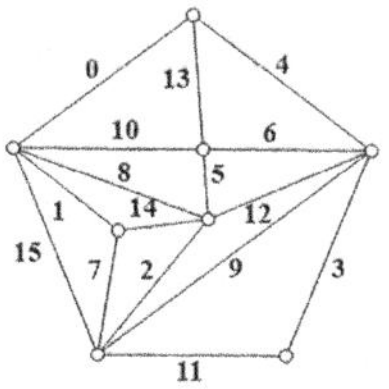


Puzzle 181

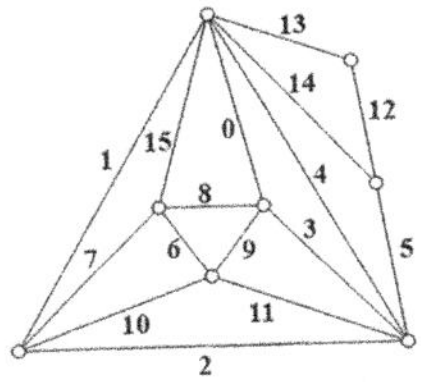


Puzzle 182

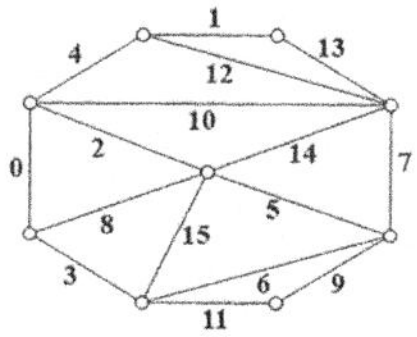


Puzzle 183

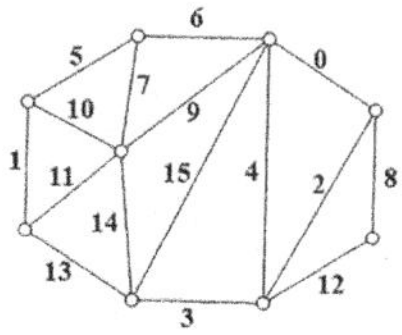


Puzzle 184

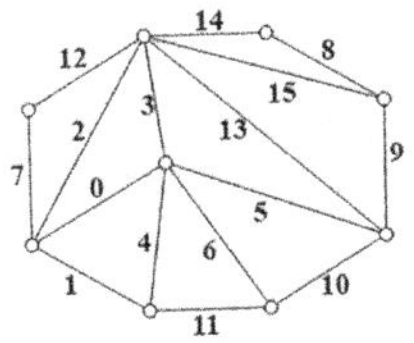


Puzzle 185

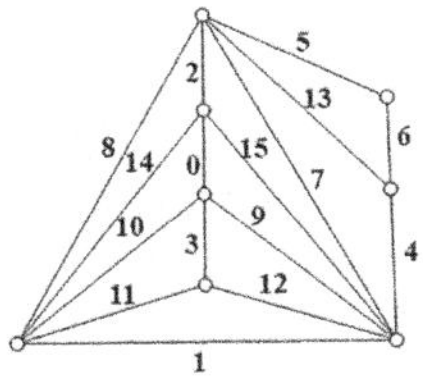


Puzzle 186

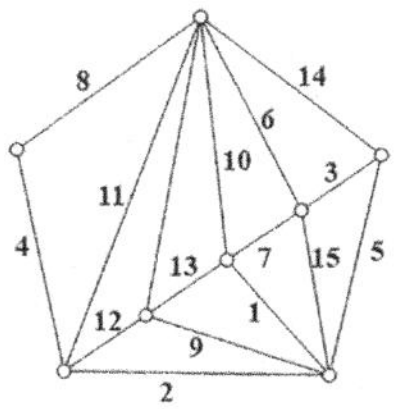


Puzzle 187

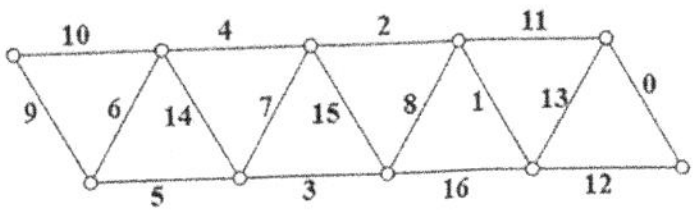


Puzzle 188

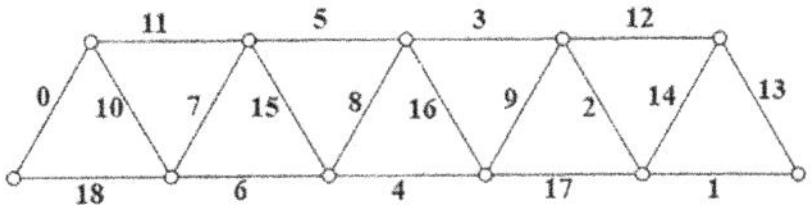

Puzzle 189

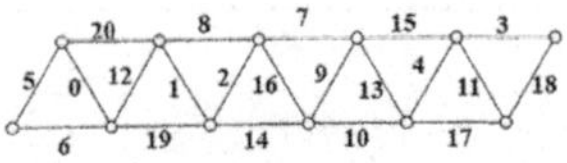


Puzzle 190

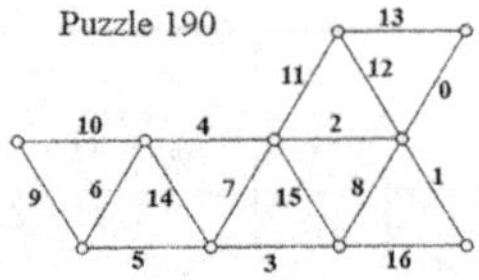


Puzzle 191

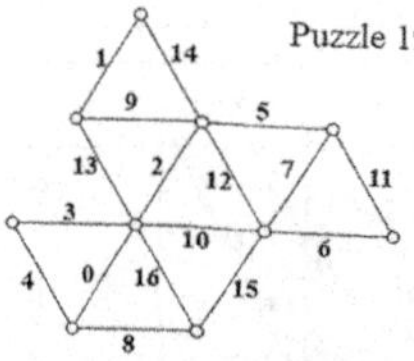


Puzzle 192

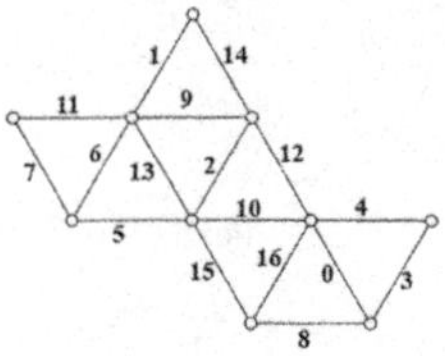


Puzzle 193

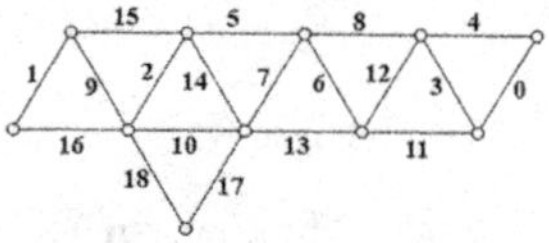


Puzzle 194

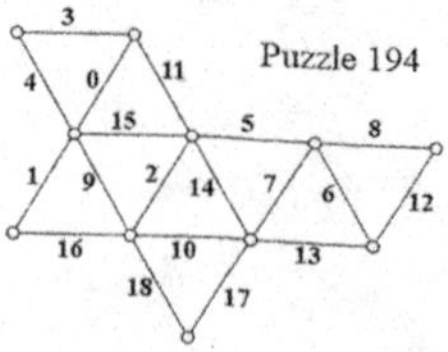


Puzzle 195

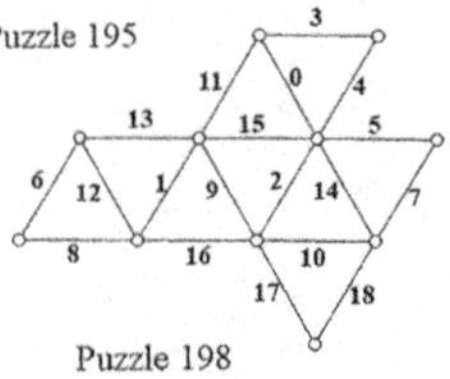


Puzzle 196

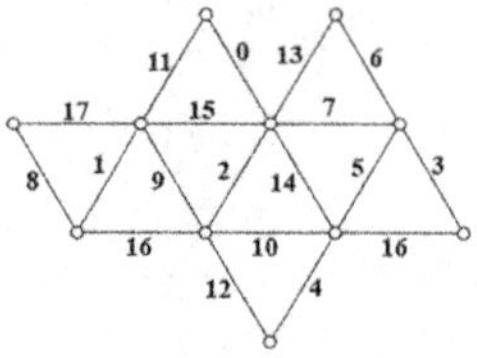


Puzzle 197

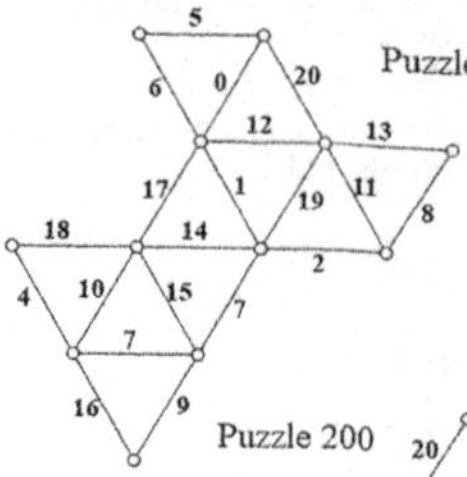


Puzzle 198

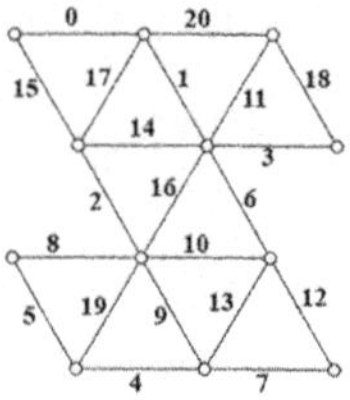


Puzzle 199

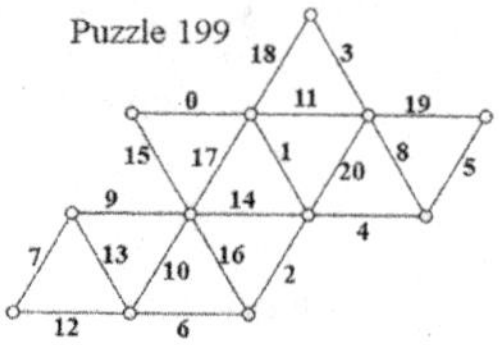


Puzzle 200

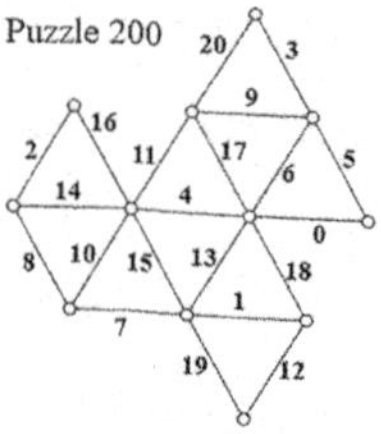


Puzzle 201

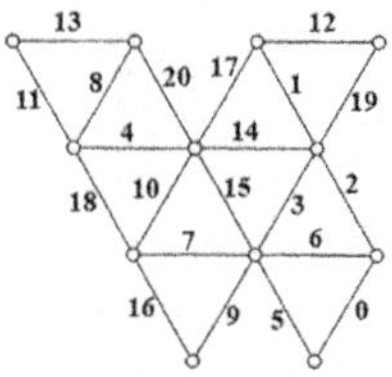


Puzzle 202

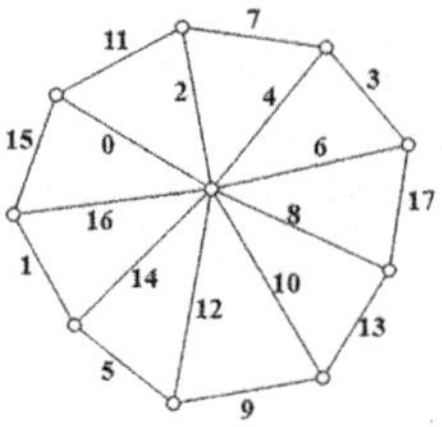


Puzzle 203

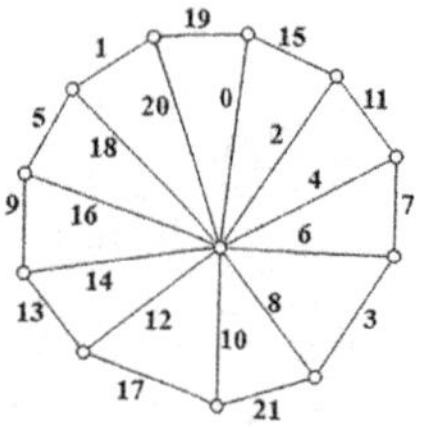

Puzzle 204

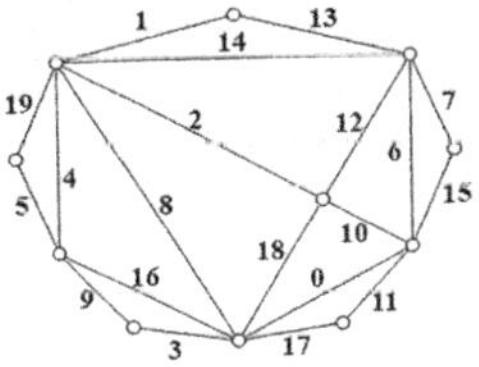


Puzzle 205

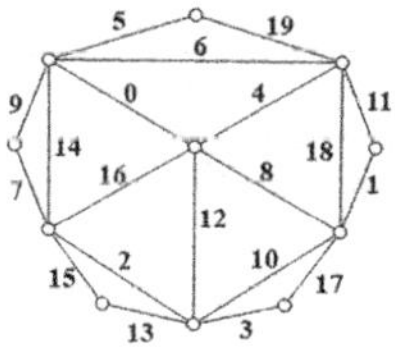


Puzzle 206

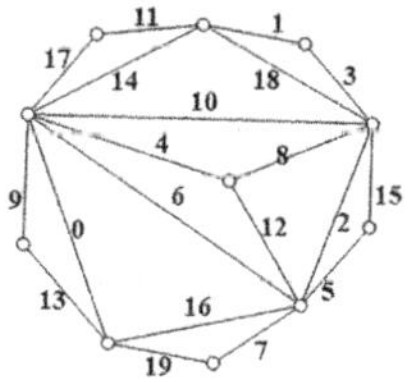


Puzzle 207

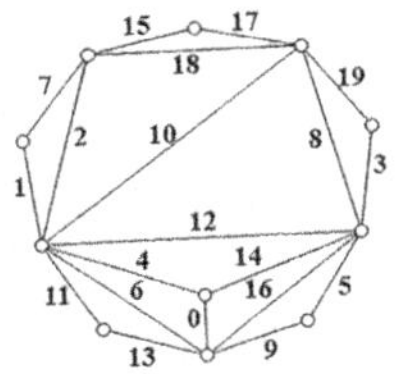


Puzzle 208

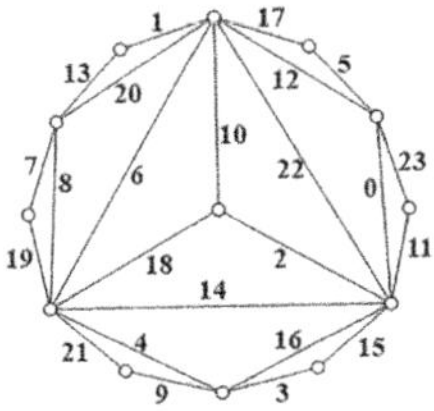


Puzzle 209

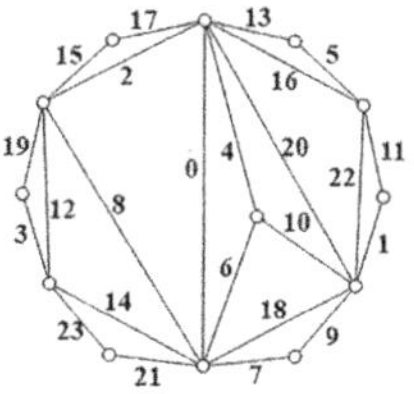


Puzzle 210

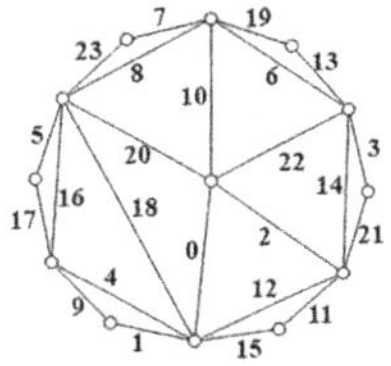


Puzzle 211

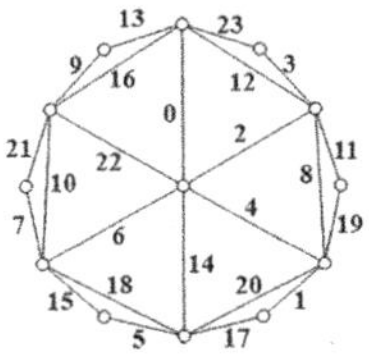


Puzzle 212

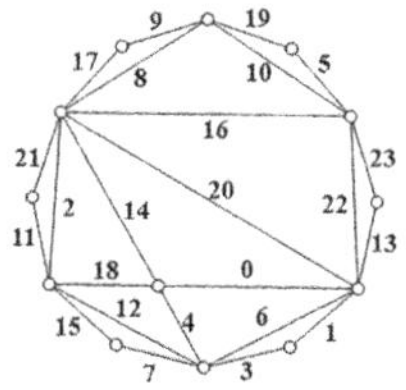


Puzzle 213

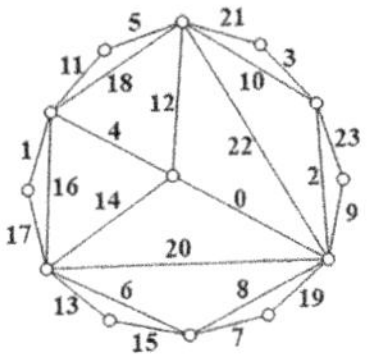


Puzzle 214

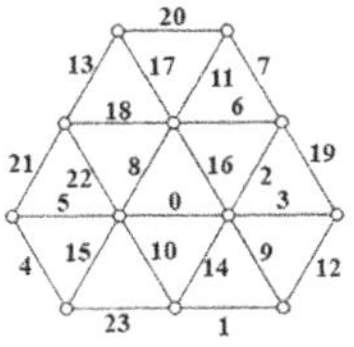


Puzzle 215

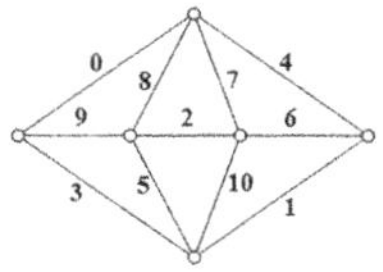


Puzzle 216

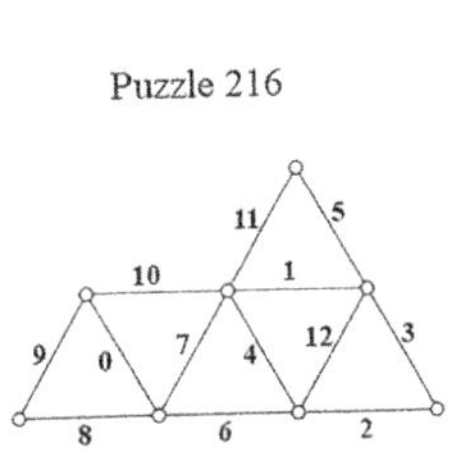


Puzzle 217

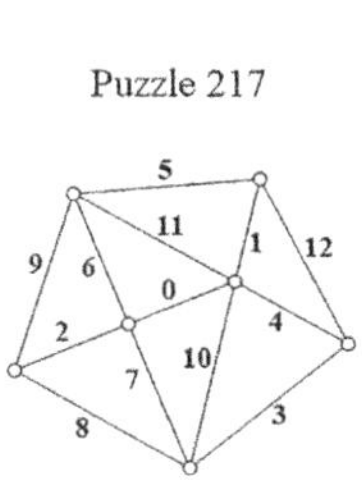


Puzzle 218

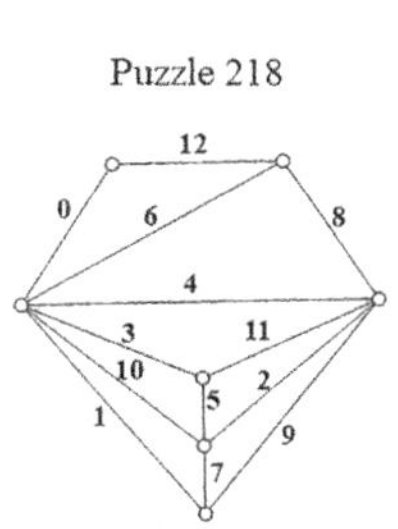

# Polyhex Puzzle Solutions

Puzzle 219

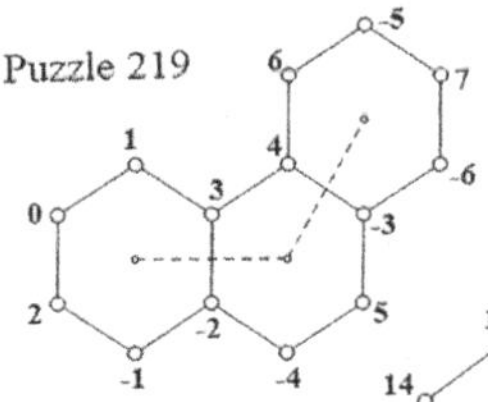


Puzzle 220

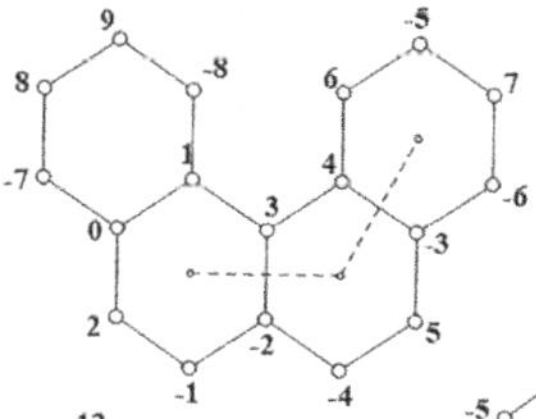


Puzzle 221

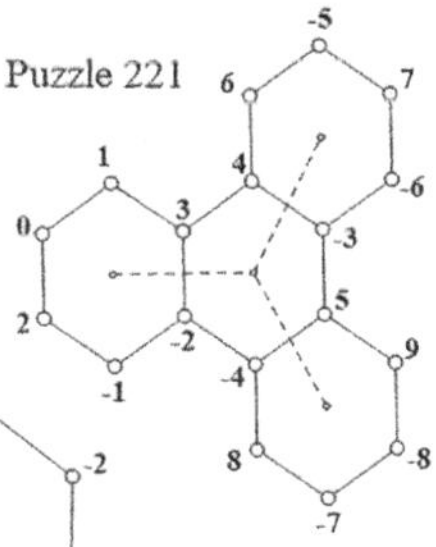


Puzzle 222

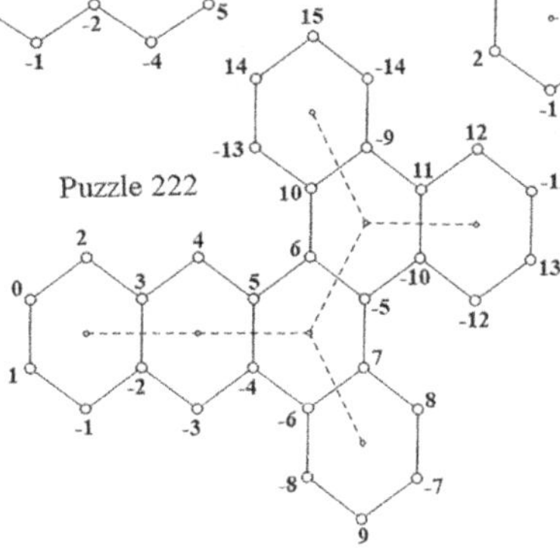


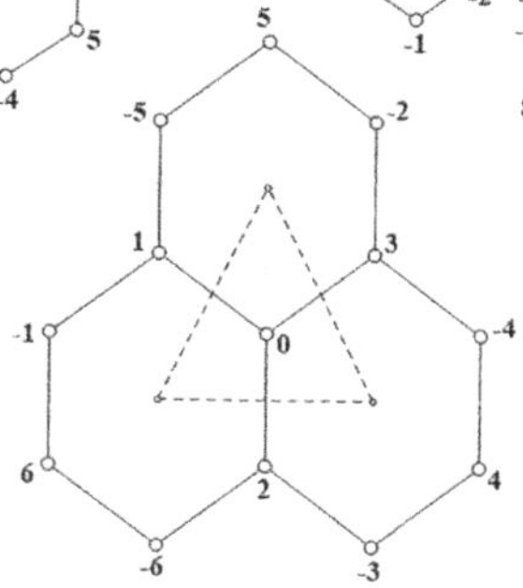


Puzzle 223

Puzzle 224

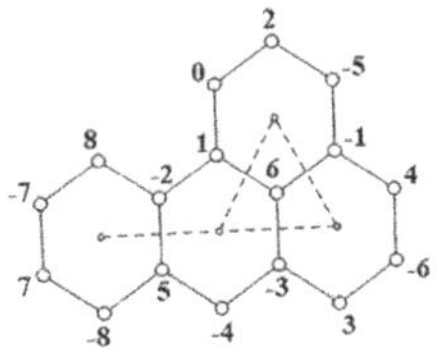


Puzzle 225

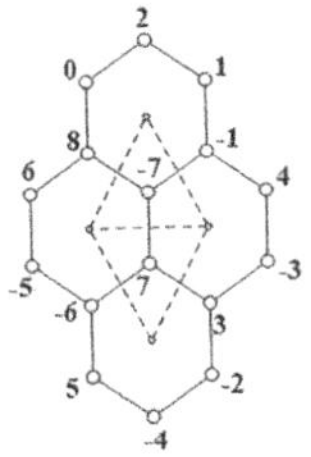


Puzzle 226

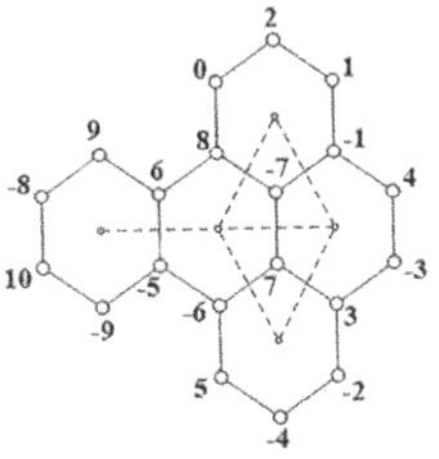


Puzzle 227

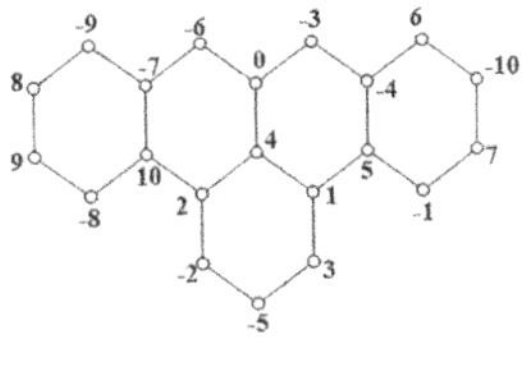


Puzzle 228

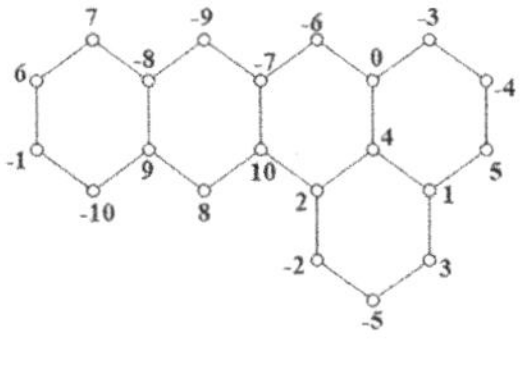


Puzzle 229

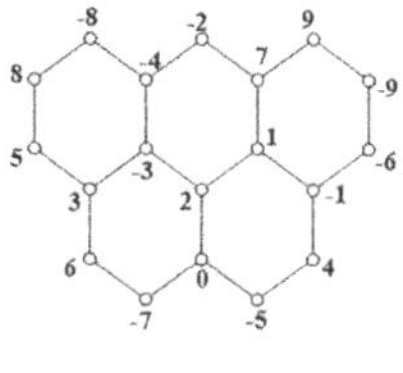

Puzzle 229

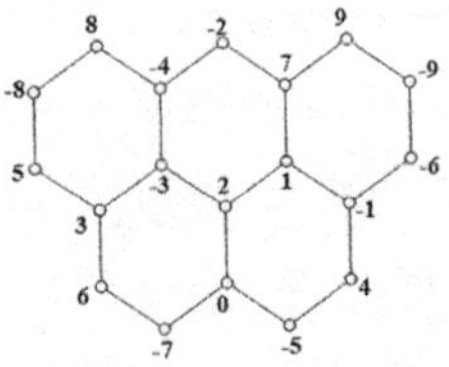


Puzzle 230

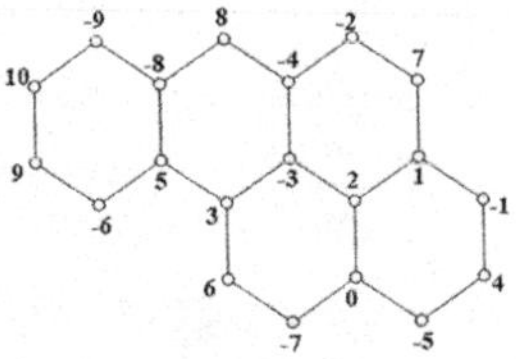


Puzzle 231

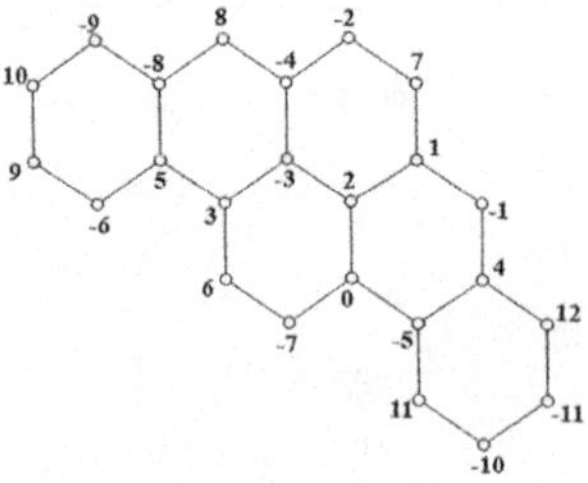


Puzzle 232

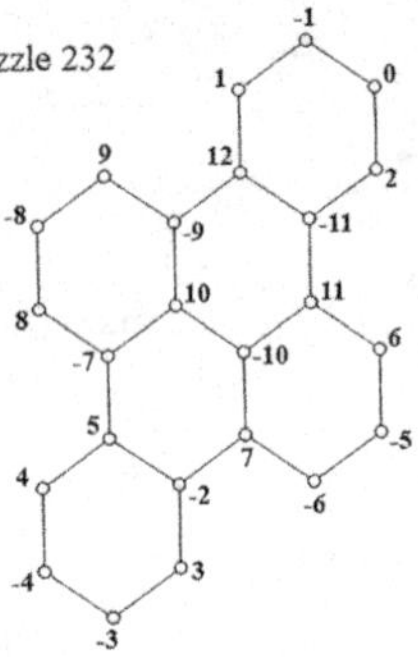


Puzzle 233

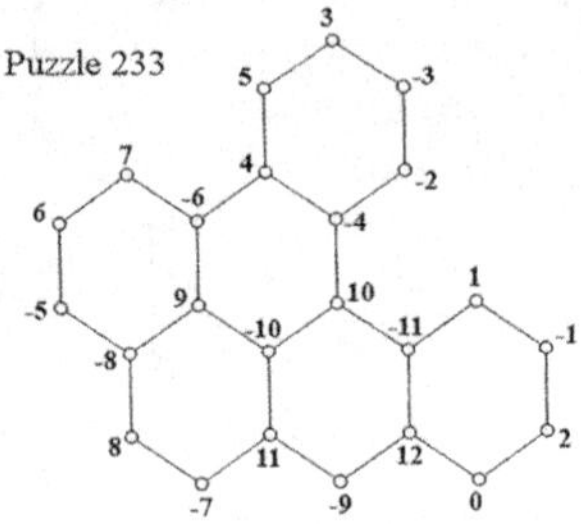


Puzzle234

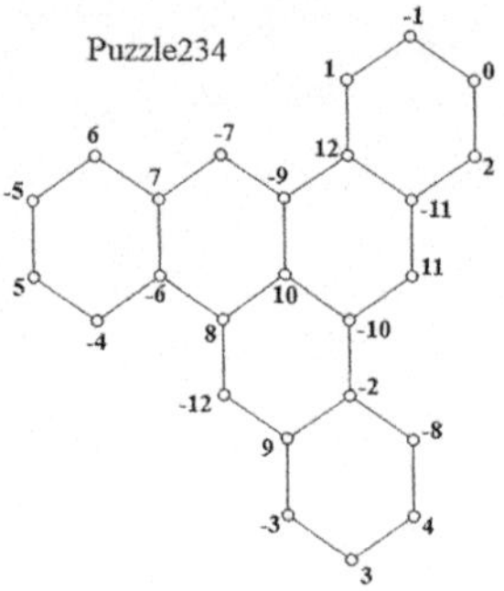


Puzzle 235

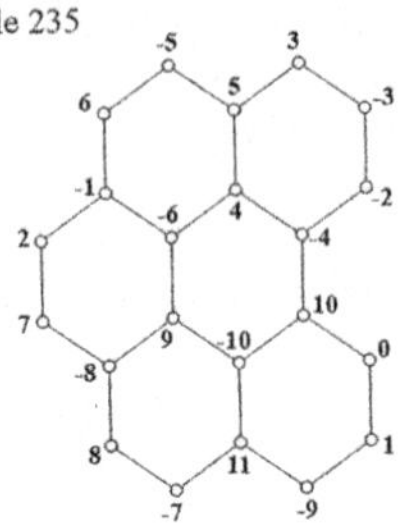


Puzzle 236

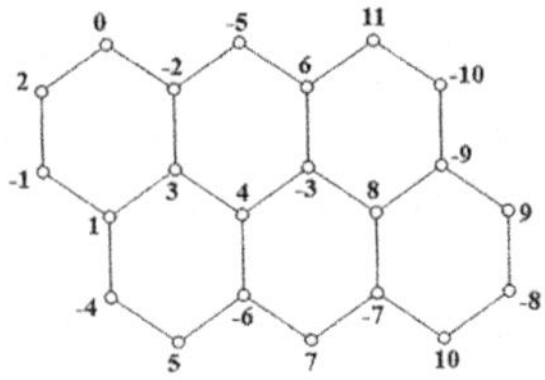


Puzzle 237

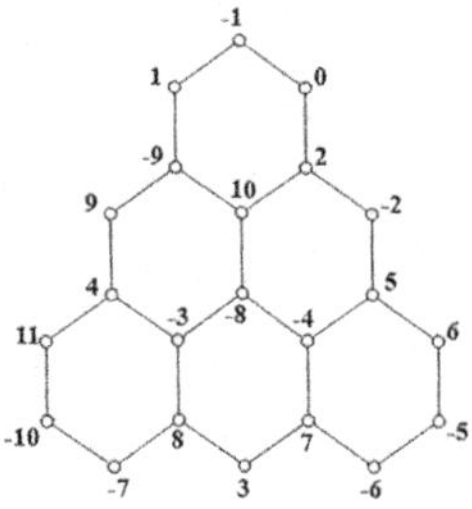


Puzzle 238

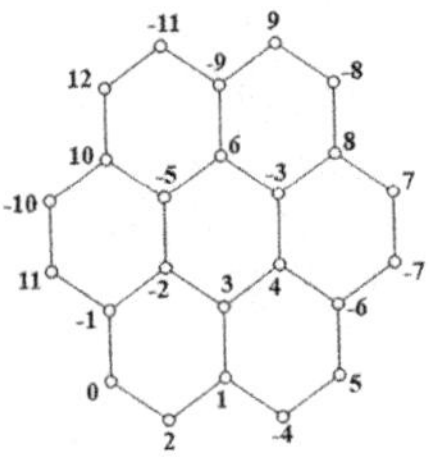


Puzzle 239

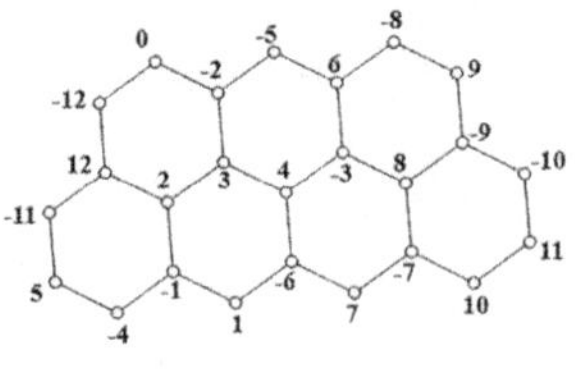


Puzzle 240

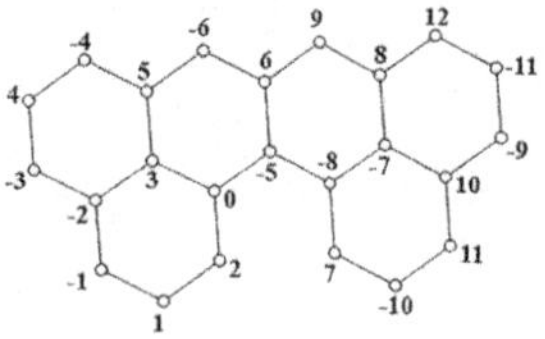

Puzzle 241

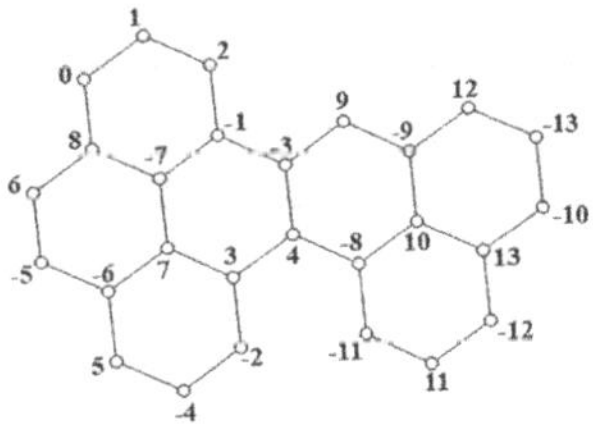


Puzzle 242

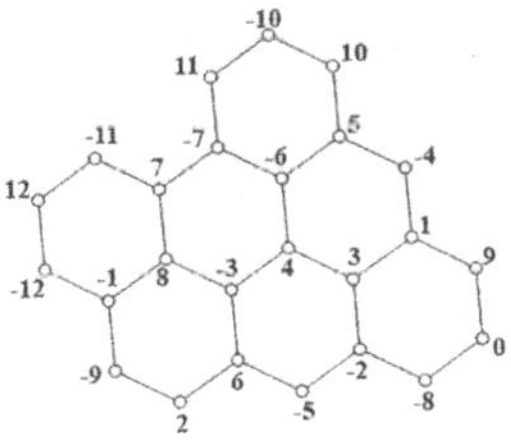


Puzzle 243

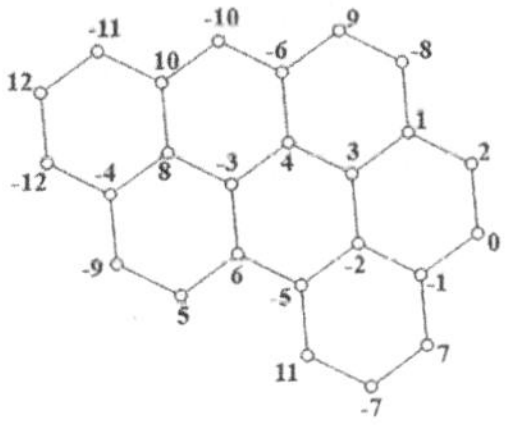


Puzzle 244

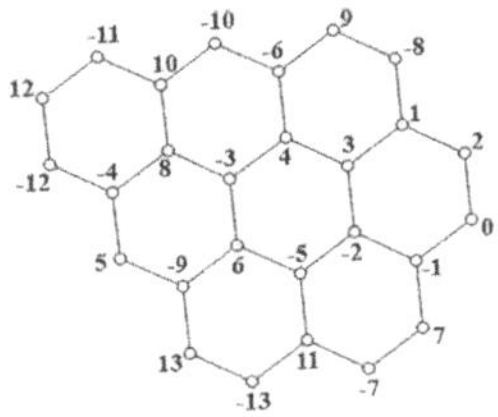


Puzzle 245

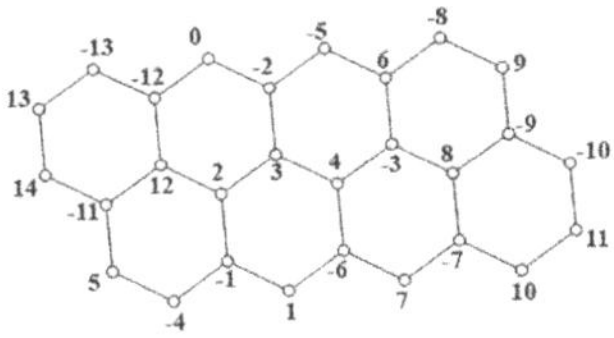


Puzzle 246

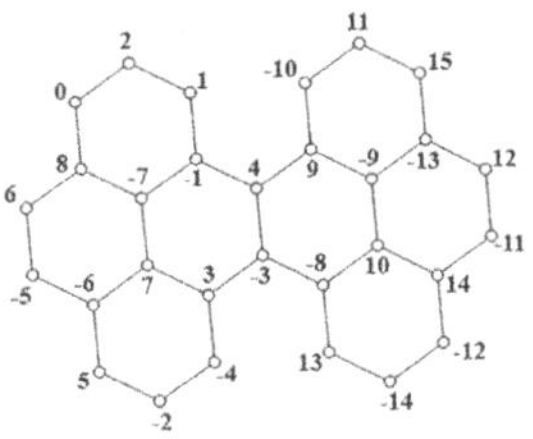


Puzzle 247

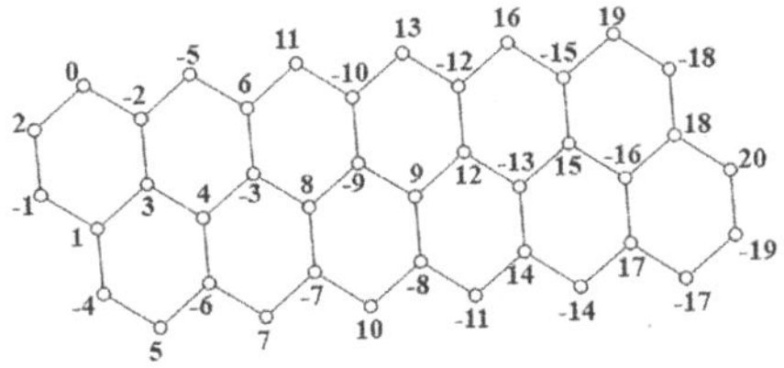


Puzzle 248

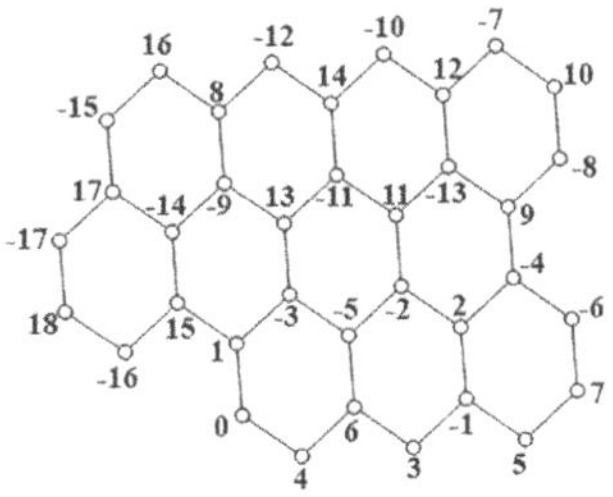


Puzzle 249

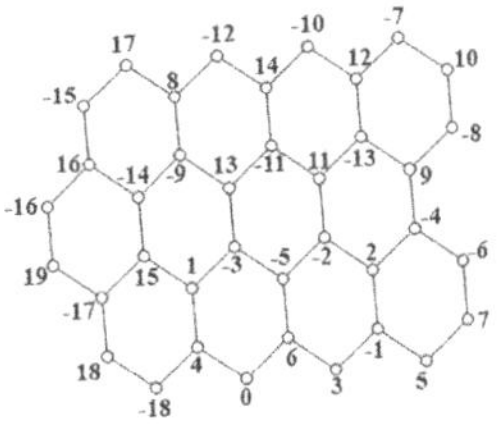


Puzzle 250

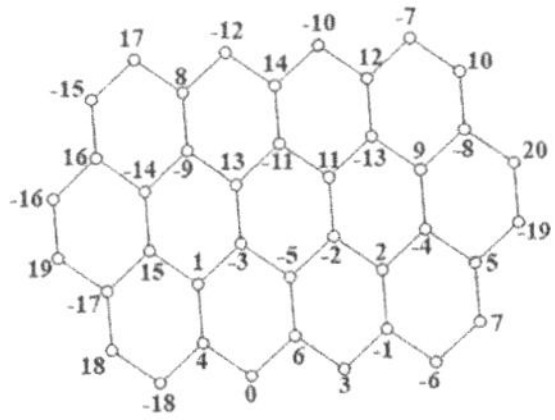


Puzzle 251

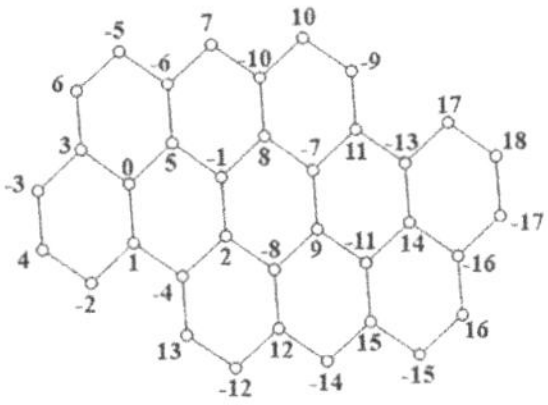

Puzzle 252

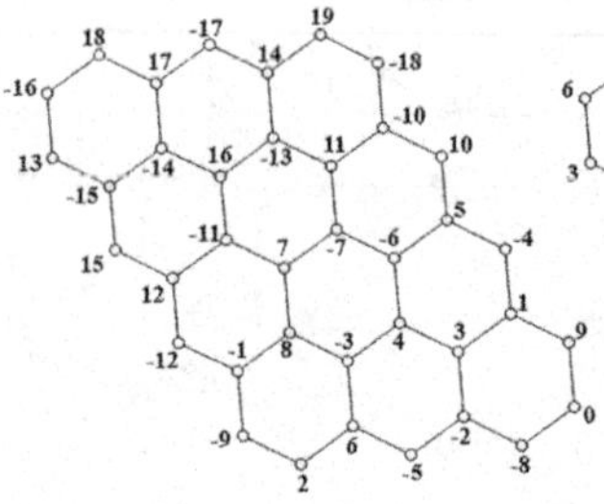


Puzzle 253

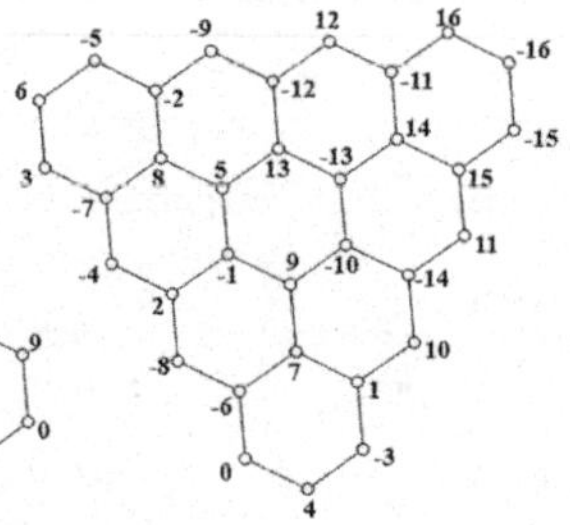


Puzzle 254

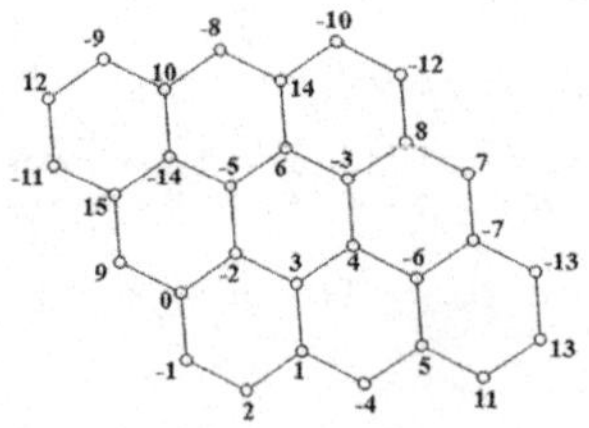


Puzzle 255

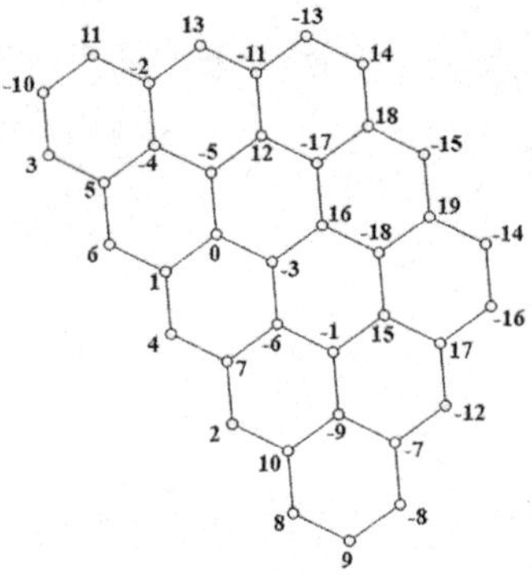


Puzzle 256

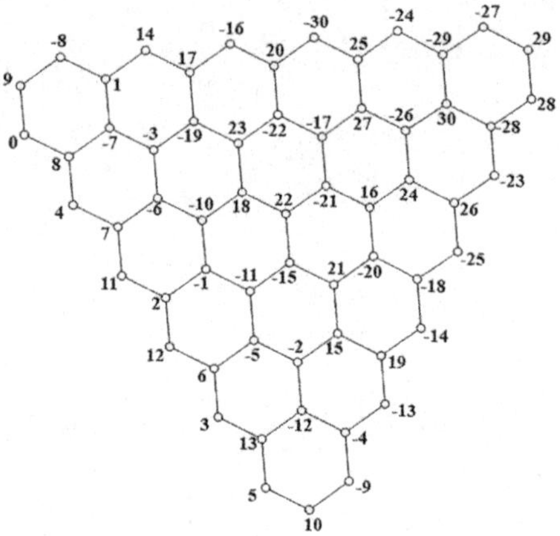

# Puzzle 257

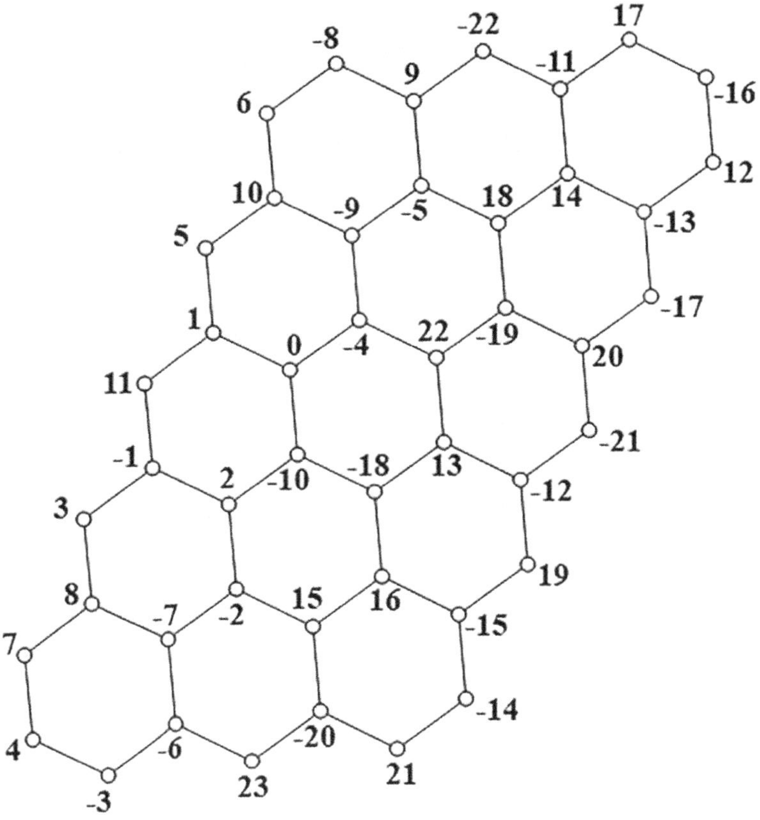
17
-22
-8
-11
9
-16
6
12
10
18
14
-9
-5
-13
5
-17
1
22
-19
0
-4
20
11
-21
-1
-18
13
2
-10
-12
3
19
8
15
16
-7
-2
-15
7
-14
-20
4
-6
21
23
-3

Made in the USA
Monee, IL
04 November 2026

67375594R00098